Terrestrial Field
Dissipation Studies

About the Cover

The photograph shows the application of a crop protection chemical to a field dissipation plot by using equipment that simulates typical application methods. The clip art cut-outs illustrate the primary themes of the symposium and emphasize the use of the study to meet specific regulatory requirements. (The photograph is adapted with permission from Aldos C. Barefoot. The adaptations, clip art, and equations were added by Kathy Kershaw.)

ACS SYMPOSIUM SERIES **842**

Terrestrial Field Dissipation Studies

Purpose, Design, and Interpretation

Ellen L. Arthur, Editor
Bayer CropScience

Aldos C. Barefoot, Editor
DuPont Crop Protection

Val E. Clay, Editor
Bayer CropScience

American Chemical Society, Washington, DC

Library of Congress Cataloging-in-Publication Data

Terrestrial field dissipation studies : purpose, design, and interpretation / Ellen L. Arthur, editor, Aldos C. Barefoot, editor, Val E. Clay, editor.

  p. cm.—(ACS symposium series ; 842)

  Includes bibliographical references and index.

  ISBN 0–8412–3769–7

  1. Pesticides. 2. Soils—Pesticide content. 3. Pesticides—Environmental aspects.

  I. Arthur, Ellen L., 1957- II. Barefoot, Aldos C., 1952- III. Clay, Val E., 1952- IV. American Chemical Society. Division of Agrochemicals. V. American Chemical Society. Meeting (220th : 2000 : Washington, D.C.). VI. Series.

SB951 .T44   2002
628.5′29—dc21   2002034156

The paper used in this publication meets the minimum requirements of American National Standard for Information Sciences—Permanence of Paper for Printed Library Materials, ANSI Z39.48–1984.

ac

Foreword

The ACS Symposium Series was first published in 1974 to provide a mechanism for publishing symposia quickly in book form. The purpose of the series is to publish timely, comprehensive books developed from ACS sponsored symposia based on current scientific research. Occasionally, books are developed from symposia sponsored by other organizations when the topic is of keen interest to the chemistry audience.

Before agreeing to publish a book, the proposed table of contents is reviewed for appropriate and comprehensive coverage and for interest to the audience. Some papers may be excluded to better focus the book; others may be added to provide comprehensiveness. When appropriate, overview or introductory chapters are added. Drafts of chapters are peer-reviewed prior to final acceptance or rejection, and manuscripts are prepared in camera-ready format.

As a rule, only original research papers and original review papers are included in the volumes. Verbatim reproductions of previously published papers are not accepted.

ACS Books Department

Contents

Indexes

Preface

The use of agricultural chemicals inevitably raises questions about the fate of the active ingredient and its degradation products in the environment as well as their effects on ecologically sensitive areas close to agricultural fields. In response to public concerns, worldwide regulatory agencies require extensive data from laboratory studies and field tests that are used to make decisions on acceptable uses of the product. In almost all countries, laboratory studies provide the primary source of data for environmental assessment. The environmental assessment culminates in studies designed to answer specific questions or to provide data that can be used directly in ecological or human health risk assessment. The requirements for field studies vary from region to region and may be required always, as in the United States and Canada, or conditionally, as in Europe. Country-specific guidelines have been developed to meet uses specific to each country's environmental assessment scheme.

Although laboratory environmental fate studies are typically designed to specifically determine degradation of an active ingredient in soil and/or water in a controlled and contained environment, field studies are conducted to determine patterns of pesticide residue dissipation under actual field conditions. A major difference in studies that examine degradation versus studies that address field dissipation is that dissipation of a pesticide is a combined result of chemical and biological processes, as well as migration processes, which include runoff, leaching, sorption, plant uptake, drift, and volatilization.

Field studies are expected to provide information on the fate of an active ingredient under typical environmental and use conditions. In general, studies have evolved to examine dissipation in soil rather than to study fate in the entire terrestrial environment. With the publication of the United States Environmental Protection Agency (EPA)/Pest Management Regulatory Agency (PMRA) proposed guidance in 1998, the design, conduct, and interpretation of the terrestrial field dissipation study was reopened.

During the past three years, the field dissipation study has been the topic of intense discussion in the North American regulatory community. Following the publication of the EPA/PMRA proposed guidance in 1998, and the subsequent Science Advisory Panel Meeting, we realized that an open scientific forum for presentations of different views of the field dissipation study was desirable as a means for seeking a common understanding among registrants, regulators, and academic and contract researchers. This book demonstrates the high level of interest in field study design and data interpretation and their use within the agrochemicals research community. The various perspectives reflected in the chapters provide valuable insights and practical experience that will enlighten the development of new regulatory guidance.

Acknowledgments

The editors offer many thanks to all contributing authors, discussion panel participants, and chapter reviewers. We are grateful for the financial support of the American Chemical Society (ACS) Agrochemicals Division and the ACS Corporation Associates, and the extraordinary cooperation of authors, reviewers, and the ACS Books Department personnel. We greatly appreciate the guidance, assistance, and patience of the ACS Books Department staff: Stacy VanDerWall, Kelly Dennis, and Margaret Brown. We thank Kathy Kershaw for the cover design. We also acknowledge Jeff Jenkins and Pat Thomson for their help in coordinating the symposium.

The editors dedicate this book to Adam Kruger, David Arthur, Susan Barefoot, and Judy Clay.

Ellen L. Arthur

Bayer CropScience
17745 South Metcalf Avenue
Stilwell, KS 66085
Ellen.Arthur@bayercropscience.com

Aldos C. Barefoot

DuPont Crop Protection,
1090 Elkton Road, P.O. Box 30
Newark, DE 19714–0030
Aldos.c.barefoot@usa.dupont.com

Val E. Clay

Bayer CropScience
8400 Hawthorn Road, P.O. Box 4913
Kansas City, MO 64120
Val.clay@bayercropscience.com

Chapter 1

A U.S. Industry Viewpoint on the Design and Use of the Terrestrial Field Dissipation Study

A. C. Barefoot[1], V. E. Clay[2], P. Hendley[3], and R. L. Jones[4]

[1]DuPont Crop Protection, 1090 Elkton Road, Newark, DE 19714–0030
[2]Bayer CropScience, 8400 Hawthorn Road, Kansas City, MO 64120
[3]Syngenta Crop Protection Inc., 410 Swing Road, Greensboro, NC 27409
[4]Bayer CropScience, 2 T. W. Alexander Drive, Research Triangle Park, NC 27709–2014

Terrestrial field dissipation studies are useful for determining the rate of dissipation of parent compounds under actual use conditions, providing information on the simultaneous formation and decline of important metabolites, and providing an indication of leaching potential and accumulation of residues in soil. No significant changes to the basic study design currently used by industry are needed to meet these study objectives. Study modules to examine such processes as leaching, volatilization from soil, metabolism and formation of bound residues, and residues in plants can be added when most of the applied material reaches the soil at application. Soil sampling can also be added to studies with foliar applications, but usually obtaining precise soil dissipation rates should not be an objective of such a study. The results of terrestrial field dissipation studies should be used in environmental risk assessments, and study designs should reflect the use of the experimental results in risk assessments.

Introduction

The terrestrial field dissipation study is an environmental fate study required for U.S. and Canadian registration of all crop protection products. This study is the most costly environmental fate study, usually of the longest duration, and often is the last study to be completed before submission of a registration package. Industry welcomes the current EPA and PMRA review of the terrestrial field dissipation study guidelines and hopes that this review will lead to an informative, cost efficient study that can be used in environmental risks assessments performed as part of the registration process.

The intent of this chapter is to provide an industry viewpoint on the purpose and design of the terrestrial field dissipation study, and the use of the terrestrial field dissipation study in environmental risk assessments performed as part of the registration process. Combining a soil dissipation study with other study modules is discussed as a means of meeting data requirements for registration of an agricultural chemical.

Purpose

Defining the purpose of the terrestrial field dissipation study is essential since the design of the study is very much dependent on the study objectives. We support a purpose generally consistent with the study design used over the past 10-20 years: The terrestrial field dissipation study should determine the rate of dissipation of the parent compound under conditions representing actual use. The study should also provide information on the simultaneous formation and dissipation of significant soil degradates. The study should be conducted in such a manner so that the focus of the study is on the soil and that potential removal processes such as runoff and erosion are minimized.

The terrestrial field dissipation study should also continue to address current secondary objectives by providing information on the potential for the accumulation of residues and potential for leaching to the extent allowed by analysis of soil residues.

The terrestrial field dissipation study should be considered a higher tier study than the laboratory studies and be used along with the results of laboratory studies when appropriate in environmental assessments.

Basic Study Elements and Design

To meet our proposed objectives for the terrestrial field dissipation study, the study should be designed to determine the dissipation of residues in soil following an application directly to soil. The study should have the following elements and basic design. Most of the design elements are based on the conclusions of the Terrestrial Field Dissipation Work Group composed of industry scientists from North America and Europe (Clay and Barefoot, 2001).

Test Substance. The study should be conducted on one representative formulated product. If one or more of the proposed formulations were a controlled release formulation known to affect the dissipation rate, then an additional study on the controlled release product would be necessary. Radiolabeled test substances are acceptable for use in studies conducted on small plots, but should not be mandatory.

Sites. The number of sites and their location will depend on the proposed uses. In general, trials at four to six locations should be conducted although this number could drop to as low as two locations for compounds with very limited geographical use. A maximum of six sites located throughout Canada, the United States and Mexico would be required for a pesticide used in a wide variety of geographical and climatic areas.

Plot size. Typical plots should be about 10 x 20 m to 10 x 40 m. When the test substance is radiolabeled, small plots of approximately 1 x 3 m may be used.

Replicates. Each location should consist of one treated plot containing at least two sub-plots and one control plot. Studies conducted for the US EPA must have at least three sub-plots; however other regulatory guidelines do not specify the number of sub-plots. Cores from each subplot may be composited (see below).

Application. The pesticide should be applied at the highest single application rate. Multiple applications are also acceptable provided that the kinetics of dissipation can be determined following the final application. The application should be made to bare ground or to a cropped plot in such a manner that a clearly defined pattern of dissipation in soil can be determined (applications to cropped plots should be made such that essentially all of the pesticide reaches the soil surface).

Immediate Post-Application Sampling. Application monitors (soil filled trays, foam pads, filter paper, petri dishes, etc.) are required to confirm the treatment rate when feasible (this is sometimes not possible with soil incorporated applications or granular applications)

Number of Soil Cores. The number of soil cores is determined by the size of the plot. For a typical large plot, fifteen to twenty soil cores should be taken at each time point. Samples from each sub-plot may be composited to give one sample at each depth increment per sub-plot.

Sampling Depth. Soil should be sampled to the depth necessary to determine 90% dissipation of the pesticide or a depth of 1 m, whichever is less.

Sampling Times. Treated plots should be sampled prior to application, immediately after treatment and at intervals (daily, weekly, monthly) depending on the rate of dissipation.

Study Duration. The study should last one year or until 90% of the parent pesticide has disappeared from the soil, whichever is less. If the pesticide is known to degrade slowly and accumulation is a concern, the study may be continued into subsequent years as necessary to define the potential for accumulation.

Analytical Method. The method of analysis must include parent pesticide and significant degradation products. Methods must be sensitive enough to quantify ten percent of the applied parent in the entire soil core. The selection of degrades to monitor for should be based on the probability that the degradate will be detected, its potential for leaching, and potential toxicological or ecotoxicological effects.

Combining Terrestrial Field Dissipation With Other Study Modules

Combining terrestrial field dissipation studies with studies on other processes such as leaching and volatilization is attractive because a more complete picture of the overall loss mechanisms is obtained. Although an attractive option and in principle a potentially satisfying scientific endeavor, a comprehensive environmental fate field study has no clear use within the U.S. or Canadian registration systems. A comprehensive study may require so many compromises in design that no objective can be met. For example, in a leaching study test conditions are chosen to maximize downward movement of water and minimize runoff, while in a runoff study test conditions are chosen to minimize downward movement of water and maximize runoff. Consequently, we suggest that different types of experiments should be considered modules that can be added as required to meet specific study objectives. Since combining too many modules can result in a study becoming unwieldy and too expensive, modules should be combined only when the study objectives and design coincide and when the process studied by a specific module is expected to be significant. Sometimes, the study designs do not coincide but some information on the specific process is needed for interpretation of the main study. In this case the module could be conducted under a design that provides the specific information needed. A potential example might be the need to have some information on soil residue levels near the soil surface in a runoff study. This could be obtained by collecting surface samples in the month or two after application rather than conducting a full terrestrial field dissipation study.

Soil Applications

Adding modules to a field dissipation study is feasible when applications of the crop protection product are made such that essentially all of the material ends up on the soil surface or incorporated into the soil. Under such conditions, the usual objective of a terrestrial field dissipation study of defining the dissipation of parent and the formation and decline of metabolites in the soil can be maintained. Potential modules that we have considered for addition are leaching, runoff, volatilization, metabolism including bound residues, and plant residues (due to uptake).

Leaching. Leaching and terrestrial field dissipation modules are often combined. For example, many ground water studies include the collection and analysis of soil cores in a manner similar to that in a terrestrial field dissipation study, although the area treated is usually larger for ground water studies than terrestrial field dissipation studies. The timing of the two studies hinders the combining of the two modules to eliminate the need for separate studies. In North America, the field dissipation study is initiated earlier in the development of a compound, since it is a required study and must be submitted in order for the compound to be evaluated by regulatory agencies.. Ground water studies are often performed as part of a conditional registration, when all of the registration studies are complete and thoroughly evaluated (and generally after management practices and uses are well defined). Both regulators and registrants would like more information on potential movement to ground water earlier in the registration process, and various options have been considered including leaching indices, computer modeling, monolithic lysimeters, and soil pore water sampling. Generally the U.S. industry has sought alternatives to the prospective ground water study, which can cost as much as $3,000,000 per site.

Ground water studies are usually triggered to provide refined exposure estimates used in risk assessments for parent or metabolites, and are the highest tier study for ground water exposure assessments. If significant residues of a parent or metabolite of interest are expected to leach to ground water, then a ground water study (including a terrestrial dissipation module) may be more appropriate than a terrestrial field dissipation study. This would result in some savings over separate ground water and terrestrial field dissipation studies by minimizing the number of sites in which the dissipation in soil is determined.

Field dissipation studies may be combined with modules to determine residues in soil pore water to provide information on potential movement of residues to ground water. Evidence of significant movement to lower depths may trigger prospective ground water studies, while insignificant movement may obviate the need for further work. Various study designs have been proposed. For example, lack of movement could be demonstrated by applying a tracer at the time of application and collecting samples from soil-suction lysimeters. Another potential approach would be to show lack of movement by applying radiolabeled product to contained soil columns located in the test plot. Soil columns would be removed at specific time points and analyzed to show the lack of radiolabeled compounds in deeper soil layers. If application of radiolabeled compounds is not possible then another approach would be to apply cold active ingredient to European-style lysimeters located at a terrestrial field dissipation test site and collect leachate from the bottom of the undisturbed soil cores. The disadvantage of all of these alternative approaches would be the lack of monitoring wells to assess the extent of movement if significant residues were found in soil-suction lysimeter or leachate samples or deeper soil layers of contained soil columns. Therefore, these study designs would generally be most useful when the registrant did not expect significant movement of residues to deeper soil layers. In those instances where significant movement was expected, the registrant would normally move directly to a prospective ground water study. However, these hybrid field dissipation/soil-pore water studies do provide a valuable addition for an intermediate study design between a field dissipation study and a prospective ground water study.

Runoff. Terrestrial field dissipation studies are rarely combined with runoff studies. This is because a requirement of conducting terrestrial field studies is that they be conducted in flat areas to minimize runoff losses so that the dissipation of the crop protection product and associated metabolites is largely due to degradation or volatilization losses. In principle, because runoff losses are usually only a few percent of the amount applied, combining the two studies would usually have little impact on the measured dissipation losses. However, the time scales of runoff studies and terrestrial field dissipation studies are usually also quite different. Most of the runoff occurs in the first couple of major rainfall events after application while the terrestrial field dissipation can last up to 18 months (less if the degradation rate of parent and all included metabolites are relatively fast). Runoff and ground water modules should not be added to the same terrestrial field dissipation study due to the incompatibility in the study conditions required for each of these modules.

Volatilization. Measurements of volatilization from the soil surface can be added to a terrestrial field dissipation study. The size of the treatment area if a volatilization module is included is normally larger than used for a terrestrial field dissipation study (although prospective ground water sites are often the ideal size). Volatilization measurements should only be required for those few compounds shown to have significant volatilization losses in the tiered laboratory schemes.

Field Metabolism. Determining the route of degradation, or metabolic pathway, under typical field conditions may be useful in environmental assessments when a compound has the potential to form numerous metabolites or when formation of bound residues is a significant mechanism of degradation. When many degradates are possible, it is important to establish which of the degradates are significant under "real-world" conditions so that the environmental fate assessment can focus on the degradates that are most likely to be present. Confirmation of the pathway or an assessment of the significant metabolites can be obtained by the application of radiolabeled compound to contained soil columns (with or without leachate collection) as part of a terrestrial field dissipation study. Soil columns are removed at specific time points, and the radioactivity is characterized in soil layers where significant radioactivity is present. The design of this type of study module is very dependent on the specific study objectives and the nature of the compounds under study. This approach can be especially useful for understanding field dissipation where aerobic and anaerobic processes operate in the surface soil.

Plant Residue Studies. Although plants may have a significant effect on the soilwater content in the root zone, usually plant uptake is not a very significant route for dissipation (although this route is the mechanism by which soil-applied compounds with systemic activity is delivered to the target) and rarely exceeds more than about 5-10 percent even for mobile compounds. Hence, the plant uptake of crop protection products will have a very small effect on the dissipation, and measurements of uptake are not necessary in field dissipation studies. Residues in the plant usually will be determined in crop residue studies that are conducted on smaller scale plots, on a shorter time scale, and at multiple sites to allow determination of pesticide residues in raw agricultural commodities

Foliar Applications

The regulatory guidelines for terrestrial field dissipation studies focus on dissipation of pesticide residues in soil and do not require studies of foliar processes such as wash-off, degradation on plant surfaces, or volatilization from plant surfaces. When a compound is applied such that a significant amount of the compound does not reach the soil immediately after application, the usual goal of a terrestrial field dissipation study of determining the rate of dissipation in soil under actual use conditions cannot be easily achieved. For example, following foliar applications, foliar wash-off can result in a continuing input of active ingredients and metabolites to the soil making it difficult to estimate the rate of disappearance. An overall rate of disappearance is also difficult to obtain because of the uncertainty associated with estimating the mass of the material on foliage on an area basis. Although collecting soil samples at specific time points may provide information that is useful for the interpretation of such studies, obtaining precise soil dissipation rates should not be an objective of a study in which the application is made to foliage. Laboratory determinations of foliar dissipation rates should be the first tier study for providing inputs to computer modeling of wash-off and degradation.

Role of Terrestrial Field Dissipation Studies in Environmental Risk Assessments.

Currently, the degradation rate determinations from the terrestrial field dissipation are not used in risk assessments conducted by the EPA during the registration process, although this study provides the best indication of the persistence of parent and metabolites under actual use conditions. Almost all estimates of exposure are calculated based on laboratory data. The fact that the degradation rate data from terrestrial field dissipation studies are not used in risk assessments is remarkable given the considerable cost and effort required for conduct and review of these studies. Therefore, many in industry question why there is such emphasis on acceptable field dissipation studies, when the study produces no direct input into registration decisions. Since the study is required under U.S. and Canadian regulations, and is routinely conducted, it is incumbent upon regulatory agencies and registrants to agree on the value of the study and develop uses for the results.

Laboratory versus Field Measurements. Laboratory and field studies provide rate data with different advantages and disadvantages. The studies are usually seen as part of a tiered scheme, with the field studies contributing to higher tier evaluations when the risk assessment must be refined. The main question is whether degradation rates from laboratory studies or field studies should be used in models for estimating exposure. Advantages of laboratory experiments are that a total mass balance is possible, there is only limited uncertainty associated with collecting samples, and the laboratory study is conducted under carefully controlled conditions. The main disadvantage is that often degradation rates are significantly different (usually slower) under laboratory conditions than for degradation rates observed in the field. This is especially true for compounds with longer persistence (presumably due to changes in the soil during incubation), those which are degraded by biological mechanisms, and for many metabolites that form during the latter part of a laboratory study when biological activity may have decreased relative to initial conditions. In laboratory studies degradation rates can be obtained quickly and usually reflect only losses due to actual degradation rather than to the combination of degradation and off-site transport.

The main advantage of measurements from well-conducted terrestrial field experiments is that they provide a good estimate of dissipation in the soil under the four to six sets of actual use conditions. The main disadvantage of these studies is that several processes may contribute to the dissipation, in addition to degradation in the soil. As a result, regulatory agenices have been reluctant to substitute field dissipation rates for laboratory degradation rates in the calculation of expected environmental concentrations. However, in many field dissipation studies, degradation may be the dominant dissipation pathway. As we previously discussed, runoff, leaching, and plant uptake will almost always be insignificant in studies conducted in accordance with current guidelines. Volatilization and photolysis, although important for some compounds, are not significant processes for many compounds, and their importance in the field may be predicted from the results of laboratory studies. In the Netherlands, a set of guidelines has been set up for assessing the suitability of results from specific field studies for use in estimating degradation rates (CTB, 1999). Where it can be shown that other dissipation processes are not significant, the measured field half-life can be regarded as a good estimate of the degradation rate. For instance, for a non-volatile compound, with a short half-life, that is unlikely to leach and where no runoff losses could occur shortly after application, a field dissipation rate may be more appropriate to incorporate into exposure models if it differs from the degradation rate determined in laboratory studies.

We recommend that registration authorities in North America consider some method of adjusting the laboratory results so that degradation rates used in exposure assessments are reflective of degradation rates under actual use conditions. In the ideal situation, when degradation rates from laboratory and field agree, the choice of laboratory or field data makes little difference in the exposure assessment. When results disagree, the choice of laboratory or field data often can result in large differences in the exposure assessment. As mentioned previously, the current policy in the U.S. is to use the laboratory data. Recently, the FOCUS group working on ground water modeling in the European Union has recommended that the choice of degradation rates must be justified by the modeler, regardless of whether it is based on laboratory or field data (FOCUS, 2000).

The question of whether to use laboratory and field degradation rates also must consider how the specific values are used, and we must recognize that equal values of laboratory and field degradation rates do not necessary mean identical inputs into risk assessments. For example, in Europe, laboratory data used in most models also includes the study temperature and soil moisture. The models then calculate a degradation rate for each time step and depth interval based on the current soil temperature and soil moisture. Because the average soil temperature in most areas of Europe is lower than the temperature used in the laboratory experiment, inputting identical degradation rates for laboratory and field data usually results in greater persistence with the laboratory data. Currently laboratory degradation rates are not corrected for temperature and soil moisture in U.S. registration procedures, but features allowing such corrections have recently been introduced into a development version of PRZM. The use of field dissipation rates in computer modeling, when the rate reflects primarily degradation, should give more realistic environmental concentrations than laboratory data for most crop protection products. Inevitability some field data will provide misleading rates of degradation and the final decision on use of field data will be dependent on the properties of the product and the conditions under which the field study was conducted.

Movement to Ground Water. Information from terrestrial field dissipation studies is typically used in two different ways to assess potential movement to ground water. The most common is to use degradation rates in soil in environmental models to predict the amount of material moving past a specific depth. That is, the rate is plugged into a model to simulate degradation in the root and vadose zone, and the model calculates the movement of soil water and determines the amounts of the crop protection product and its metabolites in soil and soil water with depth. Usually field data is not corrected for variations in soil moisture and temperature, but may be corrected for changes expected in biological transformation with depth. While the global regulatory community has not reached consensus on the use of field dissipation rates, our experience in validating models suggest that model predictions could be improved by using better data for subsurface degradation rates. Field dissipation studies provide an alternative to laboratory studies for obtaining degradation rates (used as inputs to modeling) that may represent subsurface conditions more accurately than laboratory studies conducted with surface soils.

The terrestrial field dissipation study may contribute also to an assessment of the leaching potential of the crop protection product. The field dissipation study design is sufficient to determine when a substantial portion of the applied compound moves in the soil profile. Therefore, residues in the deepest soil layer may trigger a prospective ground water study in the U.S. (Fletcher et al., 1989). However, the terrestrial field dissipation study as usually practiced does not include soil pore water or ground water sampling. Therefore, this study usually cannot conclusively eliminate the possibility of a relatively small amount of the applied material (this amount depends on the application rate, the limit of detection in the soil, and the soil water content) moving downward in the soil profile. Therefore, even when no movement is observed in terrestrial field dissipation studies, prospective ground water studies are often requested as a result of the values of laboratory measurements for degradation rates and sorption to soil. We can derive some indication of leaching from a terrestrial field dissipation study, but we must distinguish field dissipation studies from leaching studies such as prospective ground water or lysimeter studies, that are specifically designed to determine the concentration of crop protection products and their metabolites in soil water or ground water.

Movement to Surface Water. The degradation rate in soil is an input into modeling of runoff into a farm pond to determine potential exposure to aquatic plants and animals. Because runoff events near the time of application usually control the exposure estimate, degradation rates in soil usually have little impact on estimated runoff losses unless degradation rates are quite rapid (half-lives less than a few days). Therefore, usually similar results are obtained with laboratory or field degradation rates.

Estimates of Exposure by Terrestrial Organisms. Peak and time-weighted average concentrations in soil are used to assess acute and chronic exposure to soil-dwelling organisms such as earthworms. For this assessment, distinguishing between losses between the various processes is not important since the goal is to assess the disappearance from the upper layer of soil where concentrations are the highest. Therefore, the actual soil measurements from terrestrial field dissipation studies are suitable for direct use in such assessments. Peak concentrations for acute risk assessments of products with single applications will not be affected by the degradation rate. However, peak soil concentrations with multiple applications and time-weighted average concentrations may depend on the degradation rate in soil. Since concentrations in soil following multiple applications can be determined mathematically, experimental verification is usually not needed to estimate exposure to terrestrial organisms via soil.

Impacts on Following Crops. Residues remaining in the root zone can affect the growth of following crops or may be taken up into following crops. As with exposure estimates for terrestrial organisms, distinguishing between losses of the various processes is not important. Usually data from rotational crop studies are used to assess impacts on following crops. However, in some circumstances actual soil measurements from terrestrial field dissipation studies are applicable for direct use in such assessments or degradation rates based on total dissipation data could be used.

Conclusions

The primary purpose for the terrestrial field dissipation study in the current U.S., Canadian, and European regulatory schemes is determining the rate of dissipation of parent compounds under actual use conditions. Terrestrial field dissipation studies also provide information on the simultaneous formation and decline of important metabolites and can give some indication of leaching potential and accumulation of residues in soil. Terrestrial field dissipation studies designed to meet the data requirements for registration do not differ significantly from the study designs currently used by industry.

Study modules can be added to the basic design of the terrestrial field dissipation study when appropriate to study such processes as leaching, volatilization from soil, metabolism (including formation of bound residues), and residues in plants when applications are made in such a manner that results in most of the applied material immediately reaching the soil. Soil sampling (usually with limited sampling intervals) can also be added as a module to runoff studies or studies involving foliar applications but obtaining precise soil dissipation rates should normally not be an objective of the study.

The results of terrestrial field dissipation study should be used in environmental risk assessments including potential movement to ground and surface water, estimates of exposure of terrestrial organisms to residues in soil, and potential impacts on following crops.

References

1. CTB. 1999. Checklist for assessing whether a field study on pesticide persistence in soil can be used to estimate transformation rates in soil. In: Handleiding voor de Toelating van Bestrijdingsmiddelen Versie 0.1. Chapter B.4 Risico voor het milieu, II Gewasbeschermingsmiddelen, b) Uitspoeling naar het grondwater, Bijlage 3, p. 19.

2. Clay, V. and A. Barefoot. Terrestrial Field Dissipation Studies: A Review of Guidance and Guidelines, in Terrestrial Field Dissipation Studies: Purpose, Design and Interpretation, ACS Symposium Series, American Chemical Society, Washington, D. C. in press.

3. FOCUS. 2000. FOCUS Groundwater Scenarios in the EU Review of Active Substances. Report of the FOCUS Groundwater Scenarios Workgroup. EC Document Reference Sanco/321/2000 rev. 2, 202 pp.

4. Fletcher, C., S.Hong, , C. Eiden, and M. Barrett. 1989. Terrestrial Field Dissipation Studies, Standard Evaluation Procedure, Hazard Evaluation Division, Office of Pesticide Programs, US Environmental Protection Agency, Washington, D.C., 20460.

Chapter 2

Terrestrial Field Dissipation Studies: A Review of Guidance and Guidelines

A. C. Barefoot[1] and Val E. Clay[2]

[1]DuPont Crop Protection, 1090 Elkton Road, Newark, DE 19714–0030
[2]Bayer CropScience, 8400 Hawthorn Road, Kansas City, MO 64120

Field dissipation studies are a key data requirement for registration of crop protection products. Guidelines for the studies have been developed by different countries to meet uses specific to each country's environmental assessment scheme. In the past 4-5 years, guidance and guidelines have been developed for field studies as the European Union (EU) has promulgated data requirements for Annex I listing and as the US Environmental Protection Agency (EPA) and Canada's Pesticide Management Regulatory Agency (PMRA) sought to harmonize data requirements. An international industry work group sponsored by the Canadian Crop Protection Institute (CPI) and the American and European Crop Protection Associations (ACPA & ECPA) convened to respond to developments in field dissipation data requirements and propose guidance that registrants can follow to produce acceptable studies. We will review and compare guidelines from North America and Europe and present our conclusions on the applicability of field dissipation studies to environmental fate assessments.

Introduction

In October 1998, The United States Environmental Protection Agency (USEPA) Environmental Fate and Effects Division (EFED) and Health Canada Pesticide Management Regulatory Agency (PMRA) presented a proposed field dissipation study guideline to the Federal Insecticide, Fungicides, and Rodenticides Act (FIFRA) Scientific Advisory Panel. The guideline was an outgrowth of an effort conducted under auspices of a North American Free Trade Agreement (NAFTA) Technical Committee to produce a harmonized study guideline that could be adopted by the Organization for Economic Cooperation and Development (OECD). The proposed guideline enunciated the goal of providing "an integrated qualitative and quantitative environmental fate assessment which characterizes the relative importance of each route of dissipation...." (1) Registrants, while agreeing in principle with the goal of providing comprehensive environmental fate assessments, nonetheless are concerned that the effort required to carry out field studies in accordance with the proposed guideline would be disproportionate to the value of the information obtained. The scope of such a study appeared to be too large to allow the useful determination of individual mechanisms of dissipation. The proposed guideline seems likely to give an interesting set of data from which to create additional research questions, rather than data useful for exposure assessments in a regulatory context requiring decisions on use of a pesticide under typical field conditions.

In response to the presentation of the proposed guideline, industry organizations in the US, Canada and Europe formed a technical work group that has the objective of working with US EPA and PMRA to produce a guideline that would meet the needs of regulators and the regulated community globally. The workgroup, which was sponsored by ACPA, CPI and ECPA, was chartered in July 1998 and was comprised of representatives from each sponsoring organization. The work group set three tasks for itself: 1) respond to the EPA/PMRA proposed guideline, 2) develop an industry consensus on an acceptable guideline, and 3) work with regulatory agencies to produce an internationally recognized guideline. In preparation for the first task, we reviewed current and proposed guidelines from North America and Europe to determine whether there were common requirements that could serve as the basis for a harmonized guideline.

Comparison of Regulatory Guidelines for Terrestrial Field Dissipation Studies

Field dissipation studies are required by different countries depending on the expected use pattern or on the rate of degradation observed in laboratory studies. A comparison of guidelines among US, Canada, EU, Germany, and Netherlands is shown in Table 1. In the US and Canada, field studies are mandatory for sup-

Table I. Comparison of regulatory guidelines

Criterion	USEPA	Canada	EU/SETAC	Germany	Netherlands
Guideline	164-1, Field dissipation studies for terrestrial uses.	T-1-255 - 6.3 (A), Dissipation and Accumulation - Terrestrial	Part 3 Field Soil Dissipation	BBA Part IV, 4-1, section 2	Part G. Appendix G.1.3.b
When Required	Mandatory; to support the registration of an end-use product for terrestrial use.	Mandatory; to support registration in Canada	$DT50_{lab}(20\ °C) > 60$ days or $DT50_{lab}(10\ °C) > 90$ days	$DT90_{lab} > 100$ days	Additional information required on leaching risk
Test Substance	End-use product, Formulation	End-use product, Formulation	Representative pesticide formulation	Formulated product	Commercial product
Application Rate and No. of Applications	Max. recommended use rate; Max. number of applications.	Max. recommended use rate; Single or multiple as recommended per proposed time of application(s).	Representative application rate. Typically one application at the highest single rate.	Single application at highest rate	Similar to typical use instructions.

Criterion	USEPA	Canada	EU/SETAC	Germany	Netherlands
Application Monitors	Not required	Not required	Not required (application equipment must be calibrated).	Not required	Not required
Required Soil Type	Must include worst case scenario for leaching and persistence.	Depends on crop distribution and associated soil types for intended use regions.	Typical of use area.	Soils used in normal agriculture, preferably similar to German soils.	Agricultural fields where pesticide will be used.
GLP Requirement	Yes	No	Yes	Yes	Yes
Study Duration	For field crop uses: 18 months for N:164-1	Until dissipation pattern is clearly established; DT90 is reached.	Until DT90 is exceeded or max. 2 years.	Until DT90 is reached or residues plateau.	2 years
No. of Sites	At least 2	2-12 depends on intended use regions	4 sites; 2 additional if used over a wide geographical range	4-6, divided over two growing seasons	>1
No. of Plots per Site	1 treated 1 untreated control	2 treated 1 untreated control	1 treated	1	1
Plot Size	10 X 20 ft – 20 X 40 ft or small plots	2-6 m^2	100 m^2	20 m^2	

Continued on next page.

Table I. *Continued*

Criterion	USEPA	Canada	EU/SETAC	Germany	Netherlands
Slope of Plot	Less than 2%	Not specified, need to be reported.	Not specified	Not specified	Slope should be reported.
Cropped vs. Bare Ground.	Crop can be present. Bare ground usually required.	Bare ground - no vegetation allowed on the plots during study.	Pesticide should be applied to bare ground; crops may be grown following treatment.	Bare ground (fallow soil), tests on cropped plots are optional.	Bare ground or cropped, % reaching soil must be determined.
Use of Maintenance Pesticides	Round-up only.	Not specified.	Herbicides allowed that will not interfere with analysis.	Not specified	Not specified. May be used for normal development of the crop.
Tillage of Test Plot	Yes, per normal agricultural practices.	No.	Shallow tillage as needed to maintain plot.	Not specified	Tillage allowed and must be reported.
Irrigation	Yes	No, only apply irrigation if crop is normally irrigated.	Irrigation as needed to simulate average rainfall.	Irrigation allowed	Irrigation must be recorded.
Minimum No. of Cores per Interval	15	20 for large plots, 10-15 for small plots;	20	20	15 duplicate samples
Sampling Pattern	Random	Random or systematic	Systematic or ran-		Not specified

Criterion	USEPA	Canada	EU/SETAC	Germany	Netherlands
		sampling pattern	dom		
Spray Tank Solutions	Not specified	Not specified	Not specified.	Not specified	Not specified
Soil Sampling Intervals	Pre-, date of application, and immediate post-application, at intervals to determine dissipation pattern, at termination.	Sampling Interval and time to determine dissipation pattern.	Pre-application, immediate post-application (within 3 hrs), 5 additional times on a logarithmic scale.	Immediately after application, 5 additional times	Immediately post-application, 4 other times
Soil Sampling Depth	Typically 90 cm (36 inches). Shallower depths acceptable if residues are not likely to leach.	To establish appropriate core depth based on vertical distribution of pesticide residues. = 60 cm (24 inches)	Appropriate depths eg. 0-10, 10-20 cm. Deeper depths (90 cm) for mobile compounds.	20 cm minimum, greater depth samples required if leaching is likely.	DT50 reported in 0-30 cm segment
Minimum Composite Samples per Depth, per Interval	3	1	1 pooled sample/depth.	1	1 pooled sample/depth
Increment Depth per Core	A max. depth of 15 cm per increment. At least 2 depths, the lowest depth analyzed should not contain pesticide residues.	At least 2. First depth of 15 cm. Lowest depth analyzed should not contain pesticide residues.	10 cm, lowest depth section should contain no residues.	20 cm core may be sub-divided	Layer depth depends on expected mobility.

Continued on next page.

Table I. *Continued*

Criterion	USEPA	Canada	EU/SETAC	Germany	Netherlands
Field Spikes	Yes	Yes	Not specified	Not specified	Not specified
Analytes to be Included in Analysis Method	Major degradative products (> 10%) in Hydrolysis, photolysis, soil metabolism studies.	Major degradative products (> 10%) detected at any time in lab studies.	Relevant metabolites	Important metabolites	Metabolites
Analytical Detection Limit Needed	0.01 ppm	10 % of initial concentration.	As needed to determine DT90 in surface soil layers.	10% of the initial concentration or lower	<1 µg/l
Use of Tracers	Not specified	Not specified	Not specified	Not specified	Not specified for field dissipation. Required for lysimeter studies.
Report Site Location and Geography	Yes	Yes	Yes	Yes	Yes
Soil Characterization by Depth; Soil Taxonomy	Yes	Yes	Yes	Not specified, but required for evaluation	Yes, by layer or soil horizon.
Weather Data	Rainfall + irrigation (cumulative), soil temperature, air temperature, pan evaporation	Rainfall + irrigation (cumulative), soil temperature, air temperature	Weather including irrigation (air temperature or soil temperature, wind velocity)	Not specified	Air and soil temperature, precipitation, evaporation

Criterion	USEPA	Canada	EU/SETAC	Germany	Netherlands
			locity, precipitation)		
Report Field Activities During Study	Yes	Yes	Yes	Not specified	Yes
Report Depth to Groundwater	Yes	Yes	Not required	Not required	Yes
Report Cropping History on Test Plot	5 years	Not specify	At least one year	At least one year	Not required.
Report Pesticides History on Test Plot	5 years	Yes	At least one year	At least one year	Yes
Storage Stability Data	Required. Condition and length of storage must be reported.	Required. Condition and length of storage must be reported.	Stability in storage must be determined.	Not specified	Not specified.
Report Soil Moisture Content and Bulk Density	Not specified	Yes	Not specified	Not specified	Bulk density required

Continued on next page.

Table I. *Continued*

Guidance Documents

USEPA: U.S. EPA Pesticide Assessment Guidelines, Subdivision N, October 1982.
Environmental Fate - Addendum 2 - Series 164-1, 1986.
Standard Evaluation Procedure for the Terrestrial Field Dissipation study, 1989. pp 15, 18, 20.
FIFRA Accelerated Registration - Phase 3 Technical Guidance, 1989.
Pesticide Reregistration Rejection Rate Analysis, U. S. EPA 738-R-93-010, Environmental Fate, September 1993.

Canada: Environmental Chemistry and Fate Guidelines for Registration of Pesticides in Canada, 1987.
Nicholson, I. K., Environment Canada, personal communication, November 15, 1991.
Nicholson, I. K., Environment Canada, personal communication, January 31, 1994.

EPA-PMRA: Guidelines for Testing of Chemicals, Draft Proposal for a Test Guideline or Guidance Document, Terrestrial Field Dissipation Studies, Collaborative Preparation by US EPA and PMRA, September 15, 1998

BBA: Guidelines for the Official Testing of Plant Protection Products, Part IV, 4-1, Fate of Plant Protection Products in Soil-Degradation, Conversion and Metabolism.

Netherlands: Behavior of the Product and its Metabolites in Soil, Water and Air, Part G, Appendix G.1.3.a, March 1995.

porting terrestrial uses and there is a long history of using field studies in environmental fate assessments (2, 3, 4, 5, 6). The emphasis of the assessment is on accounting for the pesticide behavior under actual use conditions, and field and lab studies are integrated to give a realistic view of pesticide dissipation. The qualitative assessment of pesticide behavior is coupled with computer modeling of expected environmental concentrations to determine the acceptability of pesticide uses. In Europe, the evaluation system requires data that can be compared to specific numerical trigger values. Field studies are typically triggered by the results of laboratory studies, and may not be required at all for pesticides that degrade rapidly. The EU Commission Directive requires field studies when the DT_{50} lab (20 °C) is greater than 60 days or when the DT_{50} lab (10 °C) is greater than 90 days (7,8). The field study requirement in Germany and the Netherlands also is triggered by persistence and is linked to the potential for the pesticide or its degrades to leach (9,10). Approval for use of the pesticide in Europe may depend on the results of field studies when laboratory studies indicate the pesticide may persist in soil for more than a year (ie. $DT_{90}>1$ year and $DT_{50}>3$ months) (11). In North America and in Europe, field dissipation data are seen as better indications of potential concentrations of pesticides and their degrades in soil than laboratory studies or modeling simulations.

Purpose of Terrestrial Field Dissipation Studies

The reason for conducting the field dissipation study; whether conducted for the US, Canada, EU, Germany, Netherlands or other countries (2,3,4,5,6,7,8) is to generate data that can be used in the evaluation scheme. Although the specific features of each countries' scheme are different, the purpose of the study as indicated by national guidelines is remarkably similar. The following statement from the Canadian guidelines is representative and illustrates important aspects of the data requirement (3):

"Field studies are needed to demonstrate fate in the Canadian environment… Terrestrial studies with pesticides under use conditions are necessary to substantiate laboratory findings, particularly with respect to dissipation/accumulation, leachability and carryover of residues."

In regulatory environmental fate assessment schemes the field dissipation study is a test of the predicted behavior of the pesticide under conditions that represent typical soils and climates in which the pesticide will be found. Based on laboratory studies that investigate degradation, metabolism, and mobility in artificial environments, we develop expectations for the behavior of the pesticide. The field dissipation study may confirm the expectations or it may not. If the latter, then we face the need to explain the discrepancy. Have we missed a significant mechanism of degradation? Are the environmental conditions in the field more (or less) conducive to metabolism than lab conditions? Is movement off the field unexpectedly a large component of dissipation? For most pesticides we can propose a plausible, if speculative, explanation for the results of field dissipation studies based on our knowledge of laboratory studies. In some cases, we may need to propose hypotheses that can only be confirmed by additional experiments.

Interpretation of Results

The reliance of all the evaluation systems on rates of dissipation demonstrates the most important result of the field dissipation study – the half-life, or DT_{50} and DT_{90}, for dissipation of the pesticide in soil. The result may be used to trigger additional environmental fate studies on accumulation or long term field dissipation or may be used directly in evaluation schemes (as noted earlier) or as inputs to models. Increasingly, field studies are expected to provide rates for degradation products as well as the parent, and the studies on the extent of formation and persistence of degradates have significantly increased the number of studies needed to meet environmental fate data requirements

The field dissipation study may contribute also to an assessment of the leaching potential of the pesticide; however, the field dissipation study as usually practiced does not include soil pore water or groundwater sampling. In the deliberations of the TFD-WG, we distinguished field dissipation studies from leaching studies, such as prospective groundwater or lysimeter studies, that are specifically designed to determine the concentration of pesticide in soil water or aquifers. Nonetheless, field dissipation studies give an indication of pesticide mobility and have been used to trigger leaching studies as well as minimize concerns over leaching potential.

After examining many national guidelines, we have found common themes that can provide a basis for an internationally recognized guideline. Although the studies may play somewhat different roles in environmental assessments, each guideline indicates that the studies are expected to

- determine the rate of dissipation of the parent under field conditions,
- determine extent of formation and decline of degradation products, and
- provide an indication of leaching potential.

In addition the studies provide an picture of the degradation of a pesticide under field conditions that can be compared to expectations of its behavior formed from the results of laboratory studies. This theme has been expounded previously (12) and is the conceptual basis for study designs that include multiple field dissipation study modules (13, 14).

Guideline Comparison

The guideline comparison shown in Table 1 illustrates the requirements that are common to field dissipation studies conducted for a number of regulatory agencies. Studies are conducted usually with formulated products applied in a manner similar to typical commercial applications. The study sites must be located in areas where the chemical will be used, and environmental conditions must be representative of the use area. The number of sites varies with the specific regulatory requirement, but usually the number of sites must be sufficient to provide data over the range of possible climates or geographical areas. Field dissipation studies must be conducted for a period of time sufficient to define the degradation kinetics of the parent pesticide and establish the formation and decline of significant degradates. The studies usually are conducted on bare ground, although crops are allowed under some conditions.

Soil samples should be collected at time intervals suitable for determining the rate of degradation. For chemicals with a half-life of around 50 days, a typical sampling schedule would include soil samples taken on days 0, 3, 7, 14, 21, 30, 60, 90,120 and so on through the entire length of the study. While the sampling depth requirement is expressed differently in each guideline, the intent of each guideline is similar – samples must be collected to a depth sufficient to determine the rate of degradation in the surface soil horizon. The sampling depth must be increased for compounds that are mobile and that might leach to lower depths. A 1-m depth is usually adequate to allow determination of the chemical residues in soil. In our experience, the bulk of the chemical residues remain in the top 0-50 cm depth, and any residues that leach below that level account for a small percentage of the resides and have little effect on the calculation of rate of degradation. The performance of the analytical method must be adequate to allow the quantitation of parent chemical at a level sufficient to ensure that the amount of undetected or unquantitated chemical has a negligible effect on the accuracy of the degradation rate.

The number of soil cores taken from each treated plot must be at least 15 and many guidelines require 20. All guidelines allow soil cores to be combined by depth increment to minimize the variability and analytical costs. Compositing schemes abound, and no scheme has universal appeal or is universally applicable.

Reporting requirements are also very similar. Site characterisitics, environmental conditions, sampling methods, analytical procedures, and study activities are necessary to interpret the results and must be reported.

With the background gained from the review of existing guidelines, the work group formulated general concerns and specific comments on the EPA/PMRA proposed guideline and recommended a basic study design that will assist in harmonizing the field dissipation guideline among NAFTA and EU countries. The similarities in the guidelines and in the intent of the guidelines are obvious and more significant than the differences. Consequently, the TFD-WG proposed a basic study design that should be acceptable to meet terrestrial field dissipation data requirements of many countries around the world. We have sought to view the field dissipation data requirement within the context in which the data will be used and have developed a study design that will produce data that can be used directly in ecological risk assessments or will provide inputs to models.

Recommendation for Basic Study Design

An industry view as represented by the Terrestrial Field Dissipation Work Group

Objective

The primary objective of the field dissipation study should be to determine the rate of dissipation of a pesticide in soil at several locations representing typical use areas. Additionally, the extent of formation of its degradation products is determined, and the study should broaden the findings of the laboratory rate of dissipation studies. The field dissipation study may contribute also to an assessment of the leaching potential of the pesticide; however, our basic field dissipation study design does not include a leaching component. Field dissipation studies should be distinguished from leaching studies, such as prospective groundwater or lysimeter studies, that are specifically designed to determine the concentration of pesticide in soil water or aquifers. In focusing on dissipation rate studies, we recommend simple studies that could be conducted economically at several sites rather than complicated studies that could be carried out at only one site. We anticipate that the range of dissipation rates observed and the general assessment of leaching behavior would have immediate utility in ecological risk assessments and decisions on higher tier environmental fate studies. The basic, standard field dissipation study design would be very similar to today's study design.

The field dissipation study should be considered a higher tier study than the laboratory studies and used in environmental assessments. (See EPA Pesticide Assessment Guidelines, Environmental Fate 164-1 and Environmental Chemistry and Fate Guidelines for Registration of Pesticides in Canada, 6.3.)

Key Elements of the Terrestrial Field Dissipation Study

The following are key elements of a study that should meet the data requirements of the United States, Canada, and the European Union, as we understand the terrestrial field dissipation data needed to reach registration decisions. Additional data requirements to refine the environmental exposure from runoff, leaching, volatilization, and foliar dissipation may be warranted when the ecological risk from the pesticide appears to be unacceptable based on laboratory studies. As needed, and if possible, the additional studies may be conducted in conjunction with the basic field dissipation study. We expect that any studies that go beyond the minimum study design would be conducted at the registrant's discretion or in response to a specific data requirement.

Test substance – One representative formulated product. If one of the proposed products were a controlled release formulation known to affect dissipation rate, then an additional study on the controlled release product would be required. Radiolabeled test substances are acceptable for use in small plots, but should not be mandatory.

Sites – four to six. The number of sites and location will depend on the proposed uses. A maximum of six sites located throughout Canada, the United States and Mexico would be required for a pesticide used in a wide variety of geographical and climatic areas. Similarly, four to six sites are required in Europe depending on the geographical range of the pesticide use.

Plot size – the plot size should be dependent on the objective of the study, the characteristics of the test substance (ie. whether radiolabeled or commercial product), and the application equipment. For example, a small plot receiving a ^{14}C-labeled test substance might be 1m x 3 m; a large plot receiving a commerical product might be 10 x 20 m to 20 x 40 m.

Replicates –one treated plot containing at least two replicate sub-plots; one control plot

Application –the pesticide should be applied to bare ground at the highest single application rate or to a cropped plot such that a clearly defined pattern of dissipation in soil can be determined. Usually this will mean an application to bare ground or to bare ground beneath the crop unless the crop is turf. Post-emergent applications at an early growth stage of the crop are acceptable since the crop will not intercept a significant fraction of the applied pesticide. Multiple applications will be acceptable provided the kinetics of dissipation can be determined following the final application.

Sampling –Treatment: Application monitors are required to confirm the treatment rate.

Soil – the number of soil cores will be determined by the size of the plot. For a typical large plot, fifteen to twenty soil cores should be taken at each time point.

Samples from each sub-plot may be composited to give one sample at each depth increment per sub-plot. For example, when there are three replicate sub-plots, five cores may be taken from each plot. Each set of cores may be pooled to give three samples at each depth increment per time point. Other sampling schemes that provide information on the sampling variability are also acceptable.

Depth of sampling – Soil should be sampled to the depth necessary to determine 90% dissipation of the pesticide or a depth of 1 m whichever is less.

Sampling times – Treated plots should be sampled prior to application, immediately after treatment and at intervals (daily, weekly, monthly) depending on the expected rate of dissipation. Control plots should be sampled prior to application, at the end of the study, and at two time points during the study.

Compositing –soil cores will be divided into increments and composited by depth increment. One composite for each depth segment is the minimum requirement.

Study duration –one year or until 90% of the parent pesticide has disappeared from the soil whichever is less. If the pesticide is known to degrade slowly and accumulation is a concern, the study may be continued into subsequent years as necessary to define the potential for accumulation.

Analytical method –methods are required for the parent pesticide and degradation products of concern. Methods must be able to quantify 10% of the applied parent in the entire soil core. The selection of degradates to monitor for should be based on the probability that the degradate will be detected, its potential for leaching, and potential toxicological or ecotoxicological effects.

Conclusions

The purpose of a terrestrial field dissipation study is to determine the rate of dissipation of a pesticide and the extent to which formation of degradation products occur. Only limited indications of leaching potential can be obtained. Soil dissipation design is well-established and well-understood. Acceptable soil sampling methods are readily available.

National guidelines provide good basis for a field dissipation study guideline focussed on soil sampling. A basic study design can be derived from key elements of field dissipation studies as described in various national guidelines. The basic study is similar to the designs currently in use and can be expected to provide dissipation rates that are useful for environmental risk assessments. Separate guidelines or guidance are needed for leaching studies that can fill the gap between laboratory studies and prospective groundwater studies (US) or field leaching studies (EU).

Acknowledgments

The authors undertook the preparation of this chapter on behalf of the Terrestrial Field Dissipation Work Group, and we are grateful for the contributions of all of the members: Aldos Barefoot, Murray Belyk, Val Clay, Denise Dewar, Martin Feyerabend, Tom Gilding, Hans Guth, Olaf Kellner, Michael Leggett, John Purdy, Judy Shaw and Kim Winton. We are particularly grateful to Ellen Arthur for editorial help.

References

1. USEPA, Office of Pesticide Programs, Environmental Fate and Effects Division; and Health Canada, Pest Management Regulatory Agency; Draft Proposal for a Test Guideline or Guidance Document, Terrestrial Field Dissipation Studies, September 15, 1998.
2. U.S. EPA (1982). Pesticide Assessment Guidelines. Subdivision N, Chemistry: Environmental Fate. EPA-540/9-82-021. U.S. EPA Office of Pesticide Programs, Washington, DC.
3. Agriculture Canada, Environment Canada, and Department of Fisheries and Oceans (1987). Environmental Chemistry and Fate Guidelines for Registration of Pesticides in Canada. T-1-255. Ottawa, Canada.
4. Fletcher, C., S.Hong, , C. Eiden, and M. Barrett. (1989). Terrestrial Field Dissipation Studies, Standard Evaluation Procedure, Hazard Evaluation Division, Office of Pesticide Programs, US Environmental Protection Agency, Washington, D.C., 20460.
5. US EPA Pesticide Reregistration Rejection Rate Analysis, Environmental Fate, EPA 738-R-93-010, September 1993.
6. U.S. Code of Federal Regulations. Part 158 of Title 40, Protection of the Environment.
7. European Union Council Directive 91/414/EEC, amended by Commission Directive 95/36/EC, July 14, 1995, Annex 1, Fate and Behaviour in the Environment, 7.1.1.2.2, Field Studies.
8. SETAC-Europe, Procedures for Assessing the Environmental Fate and Ecotoxicity of Pesticides, Lynch, M.R. ed., Brussels, Belgium, 1995.
9. Biologische Bundesanstalt, Federal Republic of Germany, Guidelines for the Official Testing of Plant Protection Products, Part IV, 4-1, Fate of Plant Protection Products in Soil- Degradation, Conversion and Metabolism
10. Dutch Environmental Criteria, Behavior of the Product and its Metabolites in Soil, Water and Air, Part G, Appendix G.1.3.a, March 1995.
11. European Commission, Directorate General for Agriculture Working Document, Guidance Document on Persistence in Soil, 9188/VI/97 rev. 8, 12.07.2000.
12. Mastradone, P. J., J. Breithaupt, P. J. Hannan, J. Hetrick, A. W. Jones, R. D. Jones, R. J. Mahler, S. Syslo, and J. K. Wolf, Critical Assessment of Guidelines for Terrestrial Field Dissipation of Pesticides in Agrochemical Envi-

ronmental Fate: State of the Art, eds. Marguerite Leng, Paul Zubkoff and Elizabeth Leovey, Lewis Publishers, 1995, Chapter 9, Pages 93-98.

13. A. Barefoot, Clay, V , P. Hendley and R. Jones, A U. S. Industry Viewpoint on the Design and Use of the Terrestrial Field Dissipation Study, in Terrestrial Field Dissipation Studies: Purpose, Design and Interpretation, ACS Symposium Series, American Chemical Society, Washington, D. C..in press.

14. P. Hendley, Terrestrial Field Soil Dissipation Studies in Pesticide Environmental Risk Assessment in Terrestrial Field Dissipation Studies: Purpose, Design and Interpretation, ACS Symposium Series, American Chemical Society, Washington, D.C. in press.

Chapter 3

Design of a Terrestrial Field Leaching and Dissipation Study within the European Registration Process

B. Peters[1], M. Feyerabend[1], and B. Schmidt[2]

[1]Aventis CropScience GmbH, Industriepark Höchst, G836, D–65926
Frankfurt am Main, Germany
[2]Institut Fresenius, Im Maisel 14, D–65220 Taunusstein, Germany

A special study design was established which allowed for the monitoring of soil water and solute movement without causing significant artificial effects on the transport of conservative tracers and organic compounds. The test substance glufosinate-ammonium was applied with a rate of 2 x 800 g/ha in spring under what would be considered realistic worst-case conditions in a cool climate. The special design of this outdoor field leaching study accompanied by computer modeling proved to be a sensitive and reliable method for the evaluation of the behavior of glufosinate-ammonium and its main metabolites in soil. The results of the study showed that the potential of glufosinate-ammonium and its main metabolites for leaching to ground water is negligible.

In order to place plant protection products on the European market, an assessment of their fate and behavior in soil is required. This procedure is based on the harmonized European legal framework described in Council Directive 91/414/EEC. One aim of this directive is to avoid any residues of plant protection product active ingredients in ground water, which is reflected in the quasi-zero limit of 0.1 µg/L. To reliably assess the leaching potential of pesticides a tiered approach can be used. The first step is based on the physico-chemical data, the second includes modeling and laboratory column leaching studies, while the highest tier are outdoor lysimeter or field studies.

Several study designs are reported in the literature to assess the leaching behavior of chemicals in soil. Some of them are based on deep soil sampling (*1, 2*) which is not sensitive enough considering the European ground water limit for pesticides. Others use different types of equipment like suction samplers or piezometers to obtain ground and soil water samples (*2, 3, 4, 5, 6*). Often the results of these studies are affected by the extensive disturbance of the soil profile or are difficult to interpret or to extrapolate to other conditions.

The trial presented here was specially designed to avoid any artificial influence on the leaching behavior of the test substances as a result of the soil disturbance during instrumentation or by the sampling and measuring equipment. Horizontally installed suction samplers were used to collect soil water and equipment was installed to monitor the hydrology of the vadose zone. Accompanying computer modeling was carried out to prove the validity and the plausibility of the results, for a better process understanding and for an extrapolation of the results.

Objectives of the study

To evaluate the potential for glufosinate-ammonium and its main metabolites 3-methylphoshinico-propionic acid (MPP) and 2-methyl-phoshinico-acetic acid (MPA) to reach ground water, the factors influencing their mobility and transformation in soil were identified. Furthermore the hydrology and the physical and chemical processes occurring in the unsaturated zone were investigated. In detail, the objectives of the study were:

- to establish a reliable method to qualitatively and quantitatively assess the leaching and dissipation behavior of pesticides in soil,
- to evaluate the potential of glufosinate-ammonium and its main metabolites MPP and MPA leaching to ground water under realistic worst-case conditions,

- to identify factors that influence the movement and decline of the test substance in soil,
- to characterize the relative importance of these factors and
- to determine all relevant data necessary for accompanying computer modeling.

Experimental details

Theoretical basis

In order to scrutinize the behavior of chemicals in soil, generally three basic strategies are possible:

- to monitor residues in soil,
- to monitor residues in soil water or
- to monitor residues in ground water.

The sampling and analysis of soil water and ground water are the most sensitive methods since the LOQ (limit of quantification) in water is several orders of magnitude lower than for soil: 0.05 µg/L compared to 0.01 mg/kg for glufosinate-ammonium. Collecting mobile soil water should be preferred because soil water can be monitored in several soil horizons which allows the transport of a compound to be followed through the soil profile. Therefore the evaluation of the leaching potential is not limited to the site specific ground water depth. Also, the possible appearance of residues after application in soil water is much faster than in ground water. Furthermore, the calibration of computer models can be based on soil water concentrations at different time points in several depths. These models can be used to support process understanding, to check and confirm the plausibility of the results and to transfer the actual situation of the study to different conditions. On the other hand, sampling of ground water can give important information on dilution and dissipation effects in the saturated zone on a local or regional scale. For the kinetic evaluation of the degradation behavior of the test substance in soil and therefore, a more detailed process understanding, the determination of test substance residues in soil can provide useful additional information.

The design of a field leaching and dissipation study has to reflect realistic field conditions. Despite extensive soil and water samplings and the installation of the equipment, the disturbance of the test system should be limited. The soil structure and the hydrology should remain intact and natural. A vertical or slope installation of the equipment may create artificial preferential pathways along the shafts of the instruments and the bentonite sealing or extensive disturbance of the topsoil. Also, field maintenance according to good agricultural practice is not easily possible without removing the equipment. Therefore, the horizontal installation of suction samplers and other instruments should be preferred. To take into account the spatial variability of the test field, the study should include several plots. All data relevant for the interpretation of the results and modeling should be measured on site, including for example, weather data, hydrological data and crop data.

Test Substance

Glufosinate-ammonium (GA) [ammonium-D,L-homoalanin-4-yl-(methyl) phosphinate, IUPAC] is a non-selective herbicide currently registered for various world-wide non-crop and crop uses. As for any other plant protection product, a comprehensive knowledge of degradation and mobility properties of the active substance and its main metabolites is necessary.

- Structure:

- K_{ow}:   logP <0.1 (pH 7, 22°C)
- K_{oc} or K_{clay}:   100-1230 or 2-115
- Solubility in water: 1370 g L^{-1} (22°C)
- DT50 in soil: 3-20 d

In soil glufosinate-ammonium is degraded very fast to 3-methylphosphinico-propionic acid (MPP), subsequently to 2-methylphosphinico-acetic acid (MPA), and finally to CO_2 and bound residues.

Location and soil properties of the site

The trial was performed on a freely draining permeable sandy loam with a low organic carbon content in an area of intensive agriculture typical for maize growing in Germany.

The soil type represents a reasonable worst-case with respect to leaching and complies with the requirements of the German lysimeter guideline (7): high amount of sand, low organic carbon content. The soil properties are shown in Table I. The subsoil is mainly composed of sandy and gravely river sediments with clayey oxbow lake sediments in between. Depth to the ground water table is around 8 m. The average annual rainfall is about 650-700 mm.

Table I. Soil properties of the test site

Soil layer [m]	0.0 - 0.2	0.2 - 0.3	0.4 - 0.5	0.7 - 0.8	1.1 - 1.2	1.4 – 1.5
gravel [%]	0.5	0.1	0.1	1.2	0.4	10.7
sand [%]	57.8	72.3	82.2	83.4	71.9	30.4
silt [%]	29.2	16.6	9.7	3.6	15.7	46.9
clay [%]	12.5	11.0	8.0	11.8	12.0	12.0
pH (CaCl$_2$)	7.14	7.38	7.33	7.24	7.92	8.03
organic carbon [%]	1.0	0.9	0.4	0.1	0.1	0.6
cationic exchange capacity [meq/100 g]	10.2	9.6	5.8	3.3	5.2	6.2
bulk density [g/cm³]	1.49	1.74	1.64	1.58	1.66	1.46
classification (USDA)	sandy loam	sandy loam	loamy sand	sandy loam	sandy loam	silt loam

A map showing the exact location is presented in Figure 1.

Design of the study

The test field covers an area of 1.3 ha. The field is divided into two plots of which one is treated (Plot 1) and the other serves as an untreated (Plot 2) control. Plots 1 and 2 were divided into 3 equally sized subplots (A, B and C) for soil sampling. The remaining part of each plot (subplot D) was used for the installation of equipment for monitoring and soil water sampling. The position of major features and the site layout are depicted in Figure 2.

Figure 1. Location of the test field

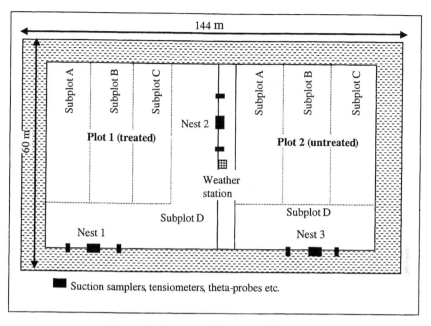

Figure 2. Design and layout of the test site

In total, three nests were established, each consisting of a central unit accommodating the batteries and the steering unit and two sub-nests. The movement of the test substance through the soil was followed by the analysis of soil water samples collected in different depths with suction samplers. These were installed at 0.3, 0.5, 0.75, 1.2 and 1.5 m in each central unit and the sub-nests. To obtain information about the soil water content and soil water tension, theta-probes and tensiometers were installed in the central units in the same depths. In order to avoid any kind of artificial preferential flow along the shafts of the equipment, the uncommon approach of fixing the whole equipment horizontally was used as shown in Figure 3.

The suction samplers were made of PVC (polyvinylchloride) fitted with a porous ceramic cup (Al_2O_3-ceramic, P80) and had a length of 1 m and 2.5-cm diameter. A polyamide tube was used for the air inlet/outlet while the sample collection tube was constructed of teflon. Since this tube ended in the ceramic cup sampled soil solution came never in contact with the PVC material and adsorption of the test substances was avoided. For installation 1-m deep holes with a diameter of 2.8 cm were augered horizontally in the soil profile from a ditch. In order to ensure a good hydraulic contact between the ceramic cup and the soil the holes were filled with a fine silica flour / water suspension before the suction samplers were fitted in the holes. The remaining void between the soil and the suction samplers was backfilled and sealed with natural soil material. Accordingly the theta-probes and the tensiometers were installed.

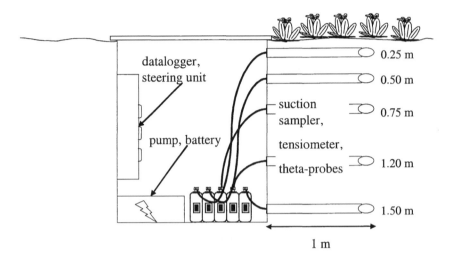

Figure 3. View on a central unit with the sampling and measuring installations

A weather station was established on site to measure air temperature, soil temperature, humidity, precipitation, global radiation, wind speed and wind direction.

A pre test was carried out with a dye tracer and LiCl to show that no artificial preferential pathways were present after the installation of the equipment.

The test substance glufosinate-ammonium was applied to a maize cropped field at the high use rate of 800 g/ha on the 20th of May 1999 and the 10th of June 1999. During the year, supplementary irrigation up to a total precipitation of > 800 mm/year was supplied to ensure crop viability and worst-case conditions. Soil water samples were taken whenever the soil water content exceeded field capacity. During the wet winter season, sampling took place twice a month. Throughout the summer the collection of soil water was initiated by strong rain events which led to water movement in the soil as indicated by the tensiometer measurements. In addition, soil samples were taken from the upper 30 cm of the field on a regular basis and divided into 10-cm segments for analysis. Finally, to investigate the waterflow in the soil, the conservative tracer KBr was applied in autumn (October 28, 1999).

Results and Discussion

The results of the pre test with a dye tracer clearly prove that there were no artificial preferential pathways present after installing all the equipment. The dye tracer needed several weeks to reach the suction samplers which can be attributed to normal chromatographic flow. By instrumenting the soil profile horizontally from a ditch, an extensive disturbance of the soil surface can be avoided. Also, problems with preferential flow along the shafts of the suction samplers or tensiometers and for example their bentonite sealing is not an issue. Furthermore there is no need to remove the equipment for field maintenance like ploughing. Nevertheless, during the study it became obvious that precautions against mouse activity were necessary.

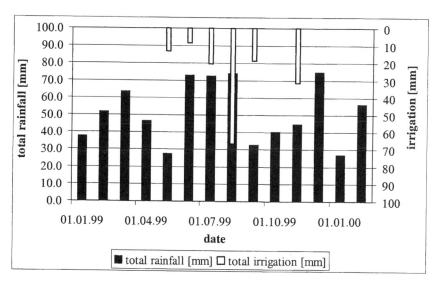

Figure 4. Rain and irrigation during the study period

In order to evaluate the study results the precipitation during the study period is depicted in Figure 4.

There were no detectable residues of glufosinate-ammonium present in the soil profile after one year. Based upon analysis of soil water samples, the results of this study after spring application show that neither glufosinate-ammonium nor its metabolites will leach below 30-cm depth in concentrations above the limit of quantification (LOQ = 0.05 µg/L). Concentrations for the 30-cm suction samplers are presented in Figure 5. Although the low sorption potential of glufosinate-ammonium and its metabolites suggests a risk for leaching of these compounds to ground water, the present study shows that their mobility in soil is limited.

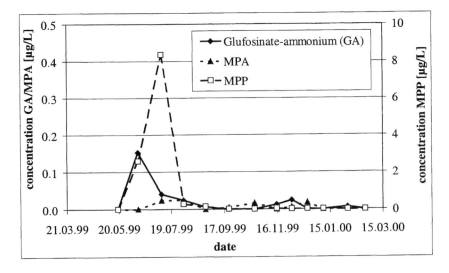

Figure 5. Residues of glufosinate-ammonium and its main metabolites in soil water (30-cm depth)

The results are even more favorable when examining the soil sample results, where with one exception glufosinate-ammonium and its metabolites were not detected below 10 cm in concentrations above the LOQ (0.01 mg/kg). After less than half a year, all residues of glufosinate-ammonium and its metabolites are below the detection limit. Figure 6 illustrates the course of the concentrations in the 0-10 cm horizon. While the first two peaks can be attributed to the two applications the reason for the third peak remained unclear. It might be caused by wash off from plant leaves.

Especially the fast degradation combined with sufficient adsorption in soil are the main reasons for glufosinate-ammonium and its metabolites not to leach to groundwater.

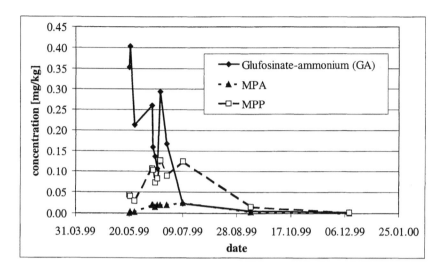

Figure 6. Residues of glufosinate-ammonium and its main metabolites in soil (10-cm depth)

Conclusion

The design of the study with the horizontal installation of the equipment proved to be appropriate to reliably assess the leaching behavior of chemicals in soil. The results of this study show that even under multiple worst-case conditions with respect to soil, precipitation and application rate, the potential of glufosinate-ammonium and its main metabolites to leach to ground water is negligible.

References

1. Jury, W. A.; Elabd, H.; Resketo M. *Water Resour. Res.* **1986**, 22(5), 749-755.
2. *Draft Proposal for a Test Guideline of Guidance Document Terrestrial Field Dissipation studies*; Draft Document September 15, 1998; U.S.

Environmental Protection Agency; Health Canada Pest Management Regulatory Agency.

3. *Gewinnung von Bodenwasserproben mit Hilfe der Saugkerzen-Methode*; DVWK Merkblätter zur Wasserwirtschaft 217; Parey, 1990.

4. Carter, A. D. *Methods of Monitoring Soil Water Regimes and the Interpretation of Data relevant to Pesticide Fate and Behaviour*; BCPC Mono. No. 47 Pesticides in Soils and Water, 1991; pp 143-147.

5. Gatzweiler, E. W.; Schmidt, B.; Feyerabend, M.; Jones, R. L.; Arnold, D. J. *Isoproturon – Field Studies on the Behaviour in Fine-Textured Soils*; Proceedings of the XI Symposium Pesticide Chemistry, Cremona; La Goliardica Pavese: Pavia, 1999; pp 305-312.

6. Jene, B. Ph.D. Thesis, SLFA Department of Ecology, Neustadt;1998.

7. *Lysimeteruntersuchungen zur Verlagerung von Pflanzenschutzmitteln in den Untergrund*; Richtlinien für die Prüfung von Pflanzenschutzmitteln im Zulassungsverfahren Teil IV, 4-3; Biologische Bundesanstalt für Land- und Forstwirtschaft, Braunschweig, 1990.

Chapter 4

Terrestrial Field Soil Dissipation Studies in Pesticide Environmental Risk Assessment

Paul Hendley

Syngenta Crop Protection Inc., 410 Swing Road, Greensboro, NC 27409

Field dissipation studies are a key data requirement for the registration of crop protection products. Guidelines for the conduct of these studies have been developed by many countries and have similar (but not quite identical) specifications. Recently a proposal to revise the United States Environmental Protection Agency (US EPA) and Canadian guidance has been put forward and promulgated as a potential Organization for Economic Cooperation and Development (OECD) guidance document; this proposal adds requirements for the studies to address the relative importance of all routes of dissipation and fully characterize the compound's dissipation in terms of "mass balance" by the incorporation of additional "modules" such as leaching and runoff. This has prompted a detailed review of the function of Terrestrial Field Soil Dissipation (TFSD) studies within regulatory risk assessment schemes from which it became apparent that, in the USA, the current TFSD study is not used significantly in risk assessment or to trigger further studies. Surprisingly, despite the high financial and reviewing costs of these studies for both regulators and industry, they provide little useful return on investment. While the proposal to design a more comprehensive study offers several advantages, a detailed review shows that, in most cases, the resulting study will not adequately address the objective of understanding mass balance. Moreover, while some modules can easily be combined with field soil dissipation elements, some of the proposed add-in "modules" are incompatible with one another as well as with the traditional conduct of field soil dissipation studies such that attempts to incorporate them all are likely to result in unproductive compromises. It is proposed that a simple guideline be developed that describes a minimum set of data to measure surface soil dissipation. The study director should then be obliged to provide additional details as needed on a case-by-case basis in order to account for other potentially important dissipation mechanisms. These modules could be added to the TFSD study if appropriate or conducted as separate studies (as is generally done at present) at the discretion of the study director. Terrestrial field soil dissipation study findings would be compared with laboratory study and exposure modeling data to identify any discrepancies that require further field, laboratory or modeling effort. Ultimately, the combined laboratory, field and modeling data sets should be used to develop an environmental fate overview that adequately describes the environmental fate of the parent material and degradates and shows that the modeled estimated exposures used for risk assessments are supported by field dissipation data to demonstrate that modeled values are "reasonable".

What is the purpose of Terrestrial Field Soil Dissipation Studies?

Before a meaningful discussion of optimal approaches for conducting Terrestrial Field Soil Dissipation (TFSD) studies on pesticides can take place, it is essential to understand their purpose. Unfortunately, this is not as straightforward as might be expected because various stakeholders in the regulatory process view the objectives somewhat differently.

Regulatory Considerations

From a traditional US EPA regulator's perspective, multiple objectives exist for the TFSD study which include two primary elements; a) to provide information on the rate at which a parent pesticide dissipates in relevant soils under relevant weather and agronomic conditions and b) to provide information on the simultaneous formation and dissipation of significant soil degradates. In addition, the study has long been regarded as the study whose results will trigger long term "build-up" and decline studies. Indeed, where it is expected that a compound or significant degradate may be persistent, the plots used for the first season of a TFSD may be laid out to serve as the physical location for a ongoing "build-up" study program. Additionally, historically, the TFSD has been considered by some to provide useful whilst limited information on the potential for transport of parent or metabolites down the soil profile.

In addition to these traditional perspectives, the recent North American proposal for draft OECD guidance for TFSD studies (1) has the stated objective to:

> "provide an integrated qualitative & quantitative environmental fate assessment which characterizes the relative importance of each route of dissipation of the parent compound and major and/or toxicologically significant transformation products"

In other words, the new draft guidance apparently substantially increases the scope of TFSD programs by requiring both qualitative and quantitative consideration of all the routes of dissipation. Presentations from some of the North American/OECD guidance document authors have suggested that this may only apply to significant dissipation routes, but this is not made clear in the current draft. Of probably even more significance for the redesign of TFSD studies is an additional stated intention that the new guidance anticipates that TFSD studies will "better address" the issue of mass balance. In other words, there is an expectation that in addition to estimating the rates of parent dissipation and the formation and decline rates of significant degradates, the study should ideally be able to quantitatively account for ultimate disposition of the mass of the originally applied chemical.

Clearly the scope of the TFSD as redefined in the recent draft guidance is now very broad. To effectively consider the value of the features added to the revised Terrestrial Field Soil Dissipation study guidance, it is first necessary to take a step backwards to develop a conceptual model of the processes that need to be accounted for. In addition it is important to understand the use made of the data generated in these studies in the pesticide regulatory process.

A Conceptual Model of Terrestrial Field Soil Dissipation

Pesticide environmental fate studies have been conducted for forty or more years and, over that period, a considerable body of technology and refined scientific approaches has been amassed. However, for the purposes of this paper, the conceptual model needs to focus on general principles only; accordingly, the author apologizes for the unavoidable over-simplifications.

Figure 1 depicts the key processes that can contribute to field dissipation. These include routes of transport away from the plot such as removal at harvest as crop residue, leaching out of the sampling zone, runoff from the plot and volatility, as well as degradative processes such as photolysis on various surfaces or biotic/abiotic degradation. While it is self-evident that for any given

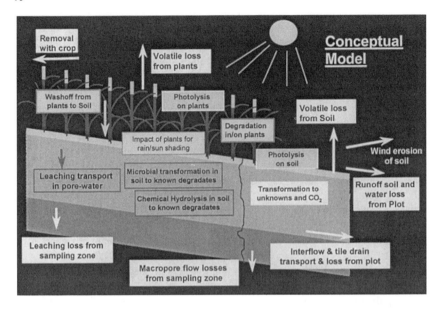

Figure 1 . Conceptual model of key processes involved in terrestrial field soil dissipation of pesticides

combination of site, crop, pesticide and use pattern, the relative importance of these processes will differ, some general principles can be derived. In particular, given the emphasis of the new guidance on mass balance accounting, it is instructive to consider how readily one might expect to be able to quantify the individual contributions of the dissipation processes to the fate of the chemical.

If the likely uncertainties associated with estimating the significance of each process are examined, it may be argued that the processes fall into three categories.

1. Those where quantification is reasonably certain, provided reasonable care is taken in making field measurements- e.g. runoff from the plot and removal of pesticide and degradates as deposits on the crop commodities and waste at harvest. Microbial and or chemical transformations in the soil to known degradates would also fall in this category.

2. Those where quantification will, even with reasonable care, be moderately uncertain. This category would include leaching below the sampling zone, volatilization from soil and/or crops and losses via tile drains.

3. Those where quantification of significance will, at best, be very imprecise. For example, transport losses via wind blown soil,

macropores or interflow or, most significantly, losses associated with degradation to unknown products/ bound residues or losses due to mineralization to CO_2.

This evaluation leads to the conclusion that completing a "mass balance accounting sheet" is a goal that will generally be unattainable.

Terrestrial Field Soil Dissipation Studies in the Current Regulatory Milieu

In the USA presently, data from a battery of laboratory studies (e.g. aerobic and anaerobic soil degradation studies, adsorption/desorption experiments, hydrolysis and photolysis degradation studies) are used to populate accepted and validated surface water and ground water exposure models [currently Pesticide Root Zone Model (PRZM) and PRZM-Exposure Analysis Model System (EXAMS)] based on conservative agronomic scenarios. The output from those models is related to laboratory toxicity information to determine if further refinement of risk is needed. Frequently, the refinement of risk is best achieved by refining the exposure estimates or by performing monitoring studies to characterize the conservatism of modeled exposure estimates. Thus additional "higher tier" field or laboratory fate studies which help to refine exposure model predictions such as foliar washoff and dissipation, runoff, field leaching (lysimeters or the Prospective Ground Water study) are triggered as a result of using exposure models which use default laboratory data. The important finding is that (with the exception of the relatively insignificant "build-up" studies) the current TFSD study is rarely, if ever, used to either trigger or rebut the need for any of the expensive higher tier studies.

What does the TFSD Study offer?

Given that the TFSD currently has little or no direct bearing on environmental or ecological risk assessment, it is important to understand what it CAN offer the registrant and regulator, given the proposed increase in experimental cost. Table I summarizes the positives and negatives associated with the current study design.

It is interesting to note that a major benefit of conducting a TFSD is to obtain field data from more than one site that helps to evaluate the "realism" of the regulatory exposure model predictions based on laboratory derived input parameters. It should be realized that this does not constitute a validation of the model since runoff and leaching are not measured in the current study design; however, it may be equally important to determine, for example, that while the use of the lab inputs predicts dissipation with a half-life of two to three weeks, the field dissipation measured values all have a half-life around 3 days. A

Table I. Summary of the positive and negative/neutral aspects associated with the current TFSD study design

Positives	Negative/Neutral
Parent Dissipation rate(s)	Information specific to site/soil/ weather conditions
Degradate field formation/ decline rate(s)	No indication of mass balance
Info on vertical movement when analyte is at concentrations above soil LOD	Limited information on leaching of compounds moving in pore water (below soil LOD)
Data to justify triggering a field build-up study	Only infrequently impacts decision to request PGW
Inferences that dissipation processes are understood	No definitive data on impact of soil texture
Field data to check realism of exposure model predictions based on Lab study inputs	Possibly inaccurate info on dissipation in presence of crop

discrepancy like this should provide a strong indication to the risk assessor that the conventional modeling is failing to account for some key dissipation process. Examples of this might be a compound which has a fast soil photolysis rate or one where anaerobic AND aerobic processes combine in the surface soil rather than the default PRZM model which assumes that only aerobic processes are significant in the surface soil.

This type of discrepancy has been at the root of the long-running and somewhat acrimonious debate between industry and EPA about whether laboratory or field dissipation data should be used as inputs for exposure models. The EPA Environmental Fate and Effects Division (EFED) guidance states that the laboratory values must be used since many complex co-occurring processes may have contributed to field half-life values. Assuming this continues to be the policy, using TFSD studies to identify discrepancies in modeling output may provide valuable justification for when it is appropriate to repeat modeling using modified input data-sets or model settings to account for additional dissipation processes.

The above discussion considers key factors relating to the existing study design; it is instructive to perform the same analysis on the new design as shown in Table II. Undoubtedly, the additional effort involved in the proposed new design generates more data and thus provides a more complete data set for examination; however, there are obvious downsides associated with making the study more complex and expensive. Perhaps the most significant conclusion is that, as explained in the consideration of the conceptual model, despite the extra cost and effort, the enhanced TFSD study design does not necessarily provide improved information accounting for the disposition and the mass balance of the applied chemicals and/or metabolites.

Can Modules Be Effectively Combined Into An All-Encompassing Field Study??

An important statement in the list of issues associated with the new TFSD design is the fact that some of the suggested "modules" in the study are not compatible within the same study design. Table III shows why this is the case. For example, while in current TFSD studies the site is managed specifically to minimize the possibilities for losses via leaching or runoff, studies designed to measure runoff must simulate "reasonable worst-case" scenarios by the use of sloping sites with impermeable soils. In contrast, prospective ground water studies must be conducted using flat sites with highly permeable soils. Therefore, to incorporate both leaching and runoff modules successfully into the new TFSD design, seriously potential site incompatibilities would have to be addressed.

However, this table does show that some modules are probably compatible within the same study design. For example, a TFSD study could successfully incorporate regular soil dissipation objectives alongside volatility and foliar washoff and foliar degradation components.

Table II. Summary of the positive and negative/neutral aspects associated with the proposed TFSD study design

Positive	Negative/Neutral
Provides understanding of interaction of more variables - may? Save money via multi-tasking	Only represents one combination of soil/weather per site
Can identify processes ignored in models & quantify their impact	Non detections in TFSD may not rebut the presumption of exposure that triggers Prospective Ground Water (PGW) or water monitoring studies.
Can help focus future studies and risk assessments	Much higher costs - more planning needed
In some cases may better address mass balance	Several "modules" incompatible
May provide a better data-set for model validation	Complexity adds GLP/technical risk and risks potential distraction by minor issues
	Runoff depends on natural rain and is unlikely to be optimal for eco-risk
	Leaching potential not optimal due to soil textures?
	Site selection to meet multiple goals may make selection of agronomically relevant sites difficult
	Data still not used for modeling
	Mass balance mostly still poor

Are All "Modules" Equally Significant to Dissipation Study Design?

Table IV examines the potential contribution that the key processes identified in the conceptual diagram are likely to make to overall dissipation of an active ingredient. It is clear that the significance of a process depends on the half life of the active ingredient; for example, for most compounds of low to moderate volatility, since volatilization tends to involve steady losses from day to day and can be weather dependent, the impact of volatile losses will increase where the stability of the compound of interest is greater.

The most important conclusion from Table IV is that based on the review by Wauchope (2), runoff of all but the longest half-life compounds (assuming the plots to be used are reasonably flat) is not likely to contribute significantly to dissipation since even under worst-case conditions the amount of compound lost from the plot will be low relative to most other processes. Therefore it would appear that it is rarely, if ever, necessary to add a "runoff module" to a field dissipation study. It is critical to realize that this is not to say that conducting separate runoff studies may not be vital to the regulatory process; for example, where the potential aquatic toxicity of a compound is significant, even small amounts of compound transport (e.g. <1%) per event might be of potential

Table III. Requirements for modules in the proposed TFSD study design

Module	Soil Permeability	Site Slope	Site activity level	Rainfall Management
Current TFSD	Typical	Flat	Low	Natural
Prospective GW Study	High	Flat	Moderate	Supplement
Runoff studies	Low	Moderate	High/Unpredictable	Natural/Supplemented
Tile Drain studies	Low with drains	Slight	Unpredictable	Natural/Supplemented
Field Volatility studies	Typical	Not relevant	High	Not relevant
Field Washoff/Degradation	Not relevant	Not relevant	Moderate	Natural/Supplemented
Proposed TFSD	Compromises?	Compromises?	Compromises?	Compromises?

Table IV. Potential contributions that the key processes identified in the conceptual diagram are likely to make to overall dissipation of an active ingredient.

Processes	Possible contribution to dissipation	Significance of each process at various soil/plant half-lives		
		Short	Moderate	Long
Biotic Degradation in Soil	0 - 100%	High	High	Moderate
Abiotic Degradation in Soil	0 - 100%	High	High	Moderate
Degradation on Plants	0 - 60%	Moderate	Moderate	Low
Total Volatile Loss*	0 - 30%	Low/Moderate	Moderate	High
Runoff from flat plot	0 - 4%	Very Low	Very Low	Very Low
Classic leaching	0 - 50%	Low	Low/Moderate	Moderate
Leaching via macropores	0 - 5%	Low	Low	Low
Washoff	No loss from plot	Low	Low	Low

* For compounds of low to moderate volatility

ecological significance such that model calibration runoff studies would be necessary.

Secondly, Table IV suggests that, for short to moderate half-life parents and degradates, a traditional soil dissipation study is all that is needed, with the possible exception of the addition of foliar degradation (removal) and/or washoff for compounds applied to crops with a high canopy coverage. However, where a compound has moderate to long soil persistence combined with significant laboratory volatility or where it has moderate to long soil persistence combined with use on lighter soils and potential to leach, it may be worth considering incorporating leaching and volatility modules or conducting separate studies to refine the understanding of the significance of leaching and volatility.

How can we improve Terrestrial Field Soil Dissipation study design??

The author recommends that EPA EFED revises the TFSD objectives to cover the following:

- Measure the field dissipation rate for parent and formation/decline rates for significant degradates under a range of realistic conditions
- Identify discrepancies by comparing model outputs based on inputs from lab studies on the compound with measured field dissipation rates
- Provide study(ies) to show model exposure estimates are "reasonable"
- Provide information to help design further studies to refine understanding and modeling of environmental behavior

In order for studies to achieve this, a basic study design should be described with simple minimum specifications. For example these might include:

- Several sites, typical soils representing the crop and/or ecoregion extent
- Bare or shaded soil; adding a cropped plot if this would improve understanding
- Sampling timing & frequency dependent on anticipated half-life
- Sampling surface soil only (0-30 cm) unless the lab half-life and adsorption data suggests otherwise

In addition to this simple "basic" TFSD study design, the study director would have the freedom to add volatility, suction lysimeters or "PGW-style" wells on an as-needed basis if desired or specified by a regulator. The driver for this would be that registrants would be obliged to provide data to address potentially significant loss mechanisms or accept the alternative of EPA's use of highly conservative assumptions in exposure assessments. These additional "modules" could be incorporated within the TFSD design where appropriate or conducted as separate modules if desired. Also, as described above, the incorporation of foliar washoff and degradation components where appropriate could conveniently be incorporated without compromising study goals.

Rather than invest a great deal of debate in devising prescriptive guidance for each type of additional "module" when incorporated into TFSD studies or conducted separately, it is proposed that a simpler approach of defining the minimum set of measurements to be reported if a study is to be regarded as acceptable.

Ensuring understanding of environmental fate.

To ensure that industry conducts all necessary studies, it is suggested that the regulators provide a strong incentive by requiring that, as part of the regulatory submission on a new active ingredient, the registrants also submit an "Environmental Fate Summary" document which draws together information from field, laboratory and exposure modeling output in order to provide a clear picture of how the compound will behave / degrade in the environment. The onus should be on the registrants to ensure that additional studies have been performed where needed to present a scientifically valid summary of the behavior and transport in the environment. Additional work might be any combination of laboratory, field or "managed field" studies. The corollary to this would be the expectation that this "Environmental Fate Summary" would receive thorough review and consideration by regulators evaluating potential exposure [as currently occurs in the European Union (EU)].

While it was shown earlier that it is probably not feasible to precisely measure the "mass balance" for many active ingredient/use pattern combinations, the proposed environmental overview document would be expected to provide a logical review of the science underlying the key loss mechanisms for the active ingredient and significant degradates, as well as to describe the expected field mass balance. This overview would also be used to explain and justify (where necessary) the selection of the parameters used as inputs for the runoff, ground and surface water exposure models. Most importantly, the findings of the terrestrial field soil dissipation study and any other "modules" would be used to demonstrate that the model output is accounting for the key environmental drivers.

References

1. USEPA, Office of Pesticide Programs, Environmental Fate and Effects Division;and Health Canada, Pest Management Regulatory Agency; Draft Proposal for a Test Guideline or Guidance Document, Terrestrial Field Dissipation Studies, September 15, **1998.**
2. Wauchope, R. D. The pesticide content of surface water draining from agricultural fields--a review. *J. Environ. Qual.* **1978**, 7, 459-472.

Chapter 5

Field Dissipation Studies: The Measurement of Zero-Time Residues

D. G. Graham[1,5], V. Clay[2], S. H. Jackson[3], and R. Jones[4]

[1]Zeneca Ag Products, 1200 South 47th Street, Richmond, CA 94804
[2]Bayer CropScience, P.O. Box 4913, 8400 Hawthorn Road,
Kansas City, MO 64120
[3]BASF, 26 Davis Drive, P.O. Box 13528,
Research Triangle Park, NC 27709
[4]Aventis CropScience, 2 T. W. Alexander Drive,
Research Triangle Park, NC 27709–2014
[5]Graham Environmental, 94 Brookfield Drive, Moraga, CA 94556

Low zero-time recoveries in field soil dissipation studies of pesticides were identified as a problem by the United States Environmental Protection Agency (EPA) in 1992[1]. Studies were designed to include various capture systems in addition to collection of standard soil cores in order to determine whether low recoveries were the result of misapplication or were from some other cause. Data from a range of capture systems for a variety of chemicals and formulation types verified that the chemical was reaching the soil. This suggested that low residues from zero-time cores were the result of sampling problems rather than application issues. Other possible contributors to low recoveries were losses after application, un-representative sampling, and losses during sample handling and preparation. Several of these processes can combine to produce low zero-time recoveries. The analysis of multiple data sets showed that low zero-time values resulted in a conservative prediction of half-life. The increase in half-life is relatively small, generally about 15%, with some up to 30%.

Background

Field soil dissipation studies are required by EPA to determine the dissipation of pesticides in the environment. Soil core samples are collected immediately after application and at subsequent intervals. An example soil-sampling plan is shown in Figure 1. As shown in this example, the plot is divided into 3 replicate sub-plots to produce samples that can be analyzed to determine variability of residues across the plot. Five cores are taken from each of the sub-plots. Zero-time samples are collected as 2.3" diameter x 6" deep cores (Figure 2) using either a hand or hydraulic metal coring tool lined with a plastic sleeve. The sleeve is removed from the metal tool and capped with plastic end caps. At that point, unless the material has been incorporated after application, all of the chemical is on the surface layer of the soil. The 5 cores taken from each of the 3 replicate sub-plots are composited to give 3 replicate samples for analysis. These samples are homogenized and sub-sampled for analysis.

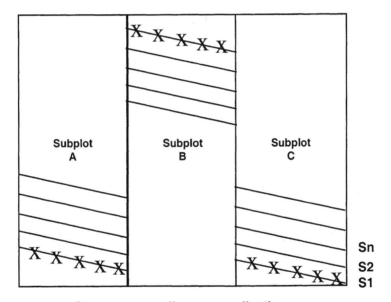

S1 = core sampling pre-application
S2 = core sampling just after application
Sn = core sampling throughout study
x = location of cores along transect

Figure 1: Example Soil Sampling Plan

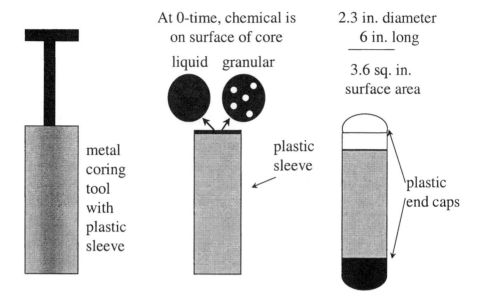

Figure 2: Soil Core Sample Collection

At issue were the low recoveries often seen from soil core samples collected immediately after application. In many cases, replicate samples gave inconsistent results, generally lower than the nominal value of application. Low results came from a wide range of formulations and application patterns. Concentrations determined from the soil core analysis increased at subsequent sampling intervals in 22% of cases. Low zero-time recovery was identified by EPA in Subdivision N Rejection Rate Analysis[2] as the number one cause of study rejection. The EPA was concerned that actual application rates were lower than required. Questions were raised about the ability of analytical methods to extract the chemical from soil samples. There was concern that half-life values derived from the studies were incorrect, possibly underestimating the length of time required for the pesticide to dissipate from the environment. And if zero-time cores are unrepresentative, what about subsequent samples?

In an effort to discover the source of these low recoveries, an American Crop Protection Association (ACPA) work group was formed and data were collected from member companies and analyzed to see if there were consistent relationships to soil type, application rate or formulation that might clarify the issue. In addition, member companies were asked to incorporate capture systems into the field trials, for comparison with results obtained from analysis of the soil cores.

Analysis of Pooled Data

Analysis of data submitted by several ACPA member companies did not show a correlation between factors such as soil type, formulation type, or

Table I. Analysis of Pooled Zero-time Recoveries - Formulation Comparison

Formulation	N	Avg % Rec.	S. Dev.	Min% Rec	Max% Rec
Granular	4	70	36	26	106
Soluble/Emulsion	115	58	32	10	133
Water dispersion	26	82	48	0	426
Not specified	39	67	26	21	107

N = number of trials for this formulation types

Avg. % Rec. = mean of recoveries from zero-time samples

S. Dev. = standard deviation

Min% Rec = minimum % recovery for this formulation type

Max% Rec = maximum % recovery for this formulation type

Table II. Analysis of Pooled Zero-time Recoveries - Application Rate Comparison

Application Rate	N	Avg % Rec.	S. Dev.	Min% Rec	Max% Rec
0.5 - 1.0	77	82	51	0	426
>1.0 - 10	86	72	36	10	222
>10	16	52	29	10	114
Not specified	5	52	30	10	114

N = number of trials with this application range

Avg. % Rec. = mean of recoveries from zero-time samples

S. Dev. = standard deviation

Min% Rec = minimum % recovery in this application range

Max% Rec = maximum % recovery in this application range

application rate and low zero-time recoveries. Data were available from 184 applications including granular, soluble/emulsion and water dispersion formulations types. Application rates ranged from 0.5 to >10 lb. a.i./acre. The data are summarized below in Tables I, II and III and presented in detail in Appendices 1 and 2.

A wide range of soil types was represented in the data sets summarized above, and there was no correlation between recovered amounts and soil texture (e.g. % sand, % silt, % clay or % organic matter).

Loss Mechanisms

Application confirmation - deposition samples

To determine whether misapplication was the real reason for the low recoveries, companies used a variety of types of targets to confirm the application. These included petrie dishes, either empty or filled with filter paper or soil, large paper targets, polyurethane foam sheets, and large pans filled with soil. The number of targets used depended on the size of plot, but generally 5 to 12 targets were used. The advantages of deposition samples include a larger surface area (up to 150 sq. inches in the case of each paper or soil tray sample as compared to 18 sq. inches for the 5 composited soil cores), less time required to secure the samples and less chance of loss because samples could be mixed or folded so that the chemical was no longer exposed on the surface of the sample. Figure 3 shows an example of target placement.

Deposition targets were placed to sample multiple spray swaths. The objective was to verify the applied amounts of chemical from different times during the application. Results are shown below in Table III. Deposition samples are better suited to measure the application than are soil cores, because they provide a more representative sample of the application and because they sample a larger area of the field. They also do not suffer from the same handling issues as zero-time soil cores

Table III: Comparison of Soil Core and Deposition Sample Recovery Data

0-day Sample Type	Number of Samples	Average Recovery, %	Standard Deviation, %
Soil Cores	93	65	25
Deposition Samples	175	83	21

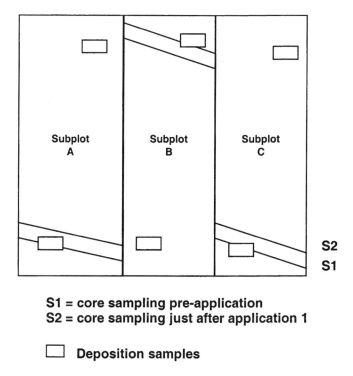

S1 = core sampling pre-application
S2 = core sampling just after application 1

Deposition samples

Figure 3: Example of Deposition Sample Placement

In another series of studies, petri dishes were used as targets. The petri dish technique used involved placing 10 g of soil in the dishes, preparing a composite sample of the soil in the three dishes immediately after application and extraction in the field. Field extraction was viewed as critical in these studies to produce consistent results for over several hundred applications. Field extraction avoided potential loss of material through contact with surfaces during shipment and subsequent handling of the samples. Results from four of these studies are shown below in Table IV.

Table IV. Comparison of petri dish and soil core recoveries

Product Type	Site	Application Rate, lb ai/acre	No. of Apps[1]	Recovery% Range Petri Dish	Recovery% Zero-Time Cores
Quinclorac	NJ-bare	0.9	2	103%	101%
herbicide	NJ-turf	0.9	2	111-113%	89%
	CA-bare	0.9	2	104-133%	64%
	CA-turf	0.9	2	88-121%	87%
	MO-bare	0.9	2	102-113%	86%
	MO-turf	0.9	2	137-143	61%
Kresoxim-methyl	NY	0.24	4	83-96%	96%
fungicide	CA	0.24	4	100-108%	88%
	OR	0.24	4	96-117%	83%
	NS	0.24	4	73-85%	78%
	ON	0.24	4	92-96%	73%
	BC	0.24	4	81-96%	134%
Tepraloxydim	ND	0.15	2	108-115%	114%
herbicide	MS	0.15	2	107-112%.	105%
Prohexidione Ca	NC	0.3	3	84-107%	106%
PGR	TX	0.3	3	100-130	69%

[1]Number of applications in trial.

Unrepresentative Samples

Several possible conditions that might lead to collection of soil cores that were not representative of the application were identified. These included:

- poor soil surface preparation - large dirt clods pushed away from narrow diameter soil probe

- fine particles lost during sample collection as dust in air venting from sampling probe
- inconsistent sample sizes - assumption of 6" sample when actual length varies

Techniques used to improve results included collecting larger diameter cores which are more representative of the field, and grooming of the soil surface with rototillers, cultivators or rollers to get a flat surface. This type of grooming may not be practical for large plots, and may actually result in problems later in the study by producing an artificially fine-grained surface that is more prone to erosion by wind or rain.

The other important step was to measure actual sample weights, so that the concentration measured in the sample could be correctly related to the surface area of the core and thus to the actual application rate measured.

Losses During Sample Handling

Several possible causes of loss during sample storage and handling have been identified. These include exposure of treated soil to plastic liner which is at a maximum in zero-time samples because all chemical is on the surface of the soil. Losses from this exposure may be either from electrostatic and lipophilic interactions - or both. Processing procedures are another source of loss. Fine particles (with a.i.) may be lost during dry ice sublimation during sample processing. Fine particles may adhere to plastic storage containers. Subsamples taken for analysis may be inhomogeneous due to inadequate mixing of homogenized sample. Properly mixing and sub-sampling samples of mixed particle size and density is difficult.

In order to minimize losses from these mechanisms, it is important to minimize storage in plastic liners. Sample preparation techniques must preserve sample integrity. In an effort to minimize the problem of getting homogeneous samples, some groups have analyzed small (i.e. 0-3") soil core sections or have analyzed each core section separately using the entire sample. This may give improved results, but at the cost of making the study more complex.

Effect of Low Zero-time Recoveries on Half-life Calculations

The effect of the zero-time value on the half-life calculated from the data collected in a field dissipation study was examined using data from 17 field dissipation studies conducted by Rhône-Poulenc (now Aventis) in the late 1980's and early 1990's. All U.S. and European trials conducted in this time period were included except for studies that involved multiple applications, granular applications, or instability of parent during analysis. These represented a wide range of chemicals and conditions including half-lives of 1-160 days, 11 different locations, different application timings, and a wide range of application rates and formulations.

In processing the data from these studies, several simplifying assumptions were made: only dissipation of the parent compound was considered, a single value (average of 4-5 replicates) expressed as percent of applied was used for each sampling interval, and sampling intervals with residue levels below quantification were discarded. Obviously, none of these simplifying assumptions are appropriate for calculations performed for regulatory use, but these assumptions allowed for uniform processing with each study represented by only 4-11 data points. For each of the 17 data sets, linear regression analyses were performed using Microsoft Excel on log transformed data with values for residues at zero-time from four different sources: the soil analyses, the filter paper analyses, and 100% and 50% of applied. Table V shows the actual values for the soil and filter paper analyses and the resulting half-life values normalized to the half-life value calculated assuming a zero-time of 100% of applied. Half-life values using zero-time values from soil and filter paper analyses ranged from 92-150% and 101-139%, respectively, of the values calculated based on a zero-time value of 100% of applied.

Table VI presents the actual half-lives and confidence intervals calculated (note: confidence intervals do not represent the values from the actual studies, since the individual replicates were not included in the statistical evaluation) assuming zero-time values of 100% and 50% of applied. The effect of lowering the zero-time value always results in an increase in the calculated half-life. Table V shows that the average increase in the calculated half-life was about 14 percent (range of 8-32 percent). The effect on the confidence interval is quite variable and depends on the relative position of the zero-time value and the calculated y-intercept.

This simple exercise demonstrated that the effect of using low zero-time values is to produce a more conservative estimate of the half-life calculated from linear regression. The effect of low zero-time values on non-linear regression is less straightforward. Almost always the lowering of the starting value will also lengthen the DT50 and DT90 values, thereby exaggerating the environmental persistence of the compound. Due to the shape of the curve sometimes the DT90 value will be lengthened quite significantly.

Table V. **Values of zero-time soil and filter paper analyses and calculated half-lives normalized to the half-life calculated using a zero-time value of 100% and 50% of applied**

Study	Measured Values (percent of applied)		Normalized Half-Life with the Indicated Zero-Time Value			
	Soil	Filter Paper	Soil	Filter Paper	100 % of Applied	50 % of Applied
1	72.6	50	105	111	100	111
2	109	78	98	105	100	117
3	111	95	99	101	100	110
4	108	80	99	103	100	108
5	64.3	NM	105	-	100	109
6	107	NM	100	-	100	122
7	163	85	92	102	100	115
8	96.7	NM	101	-	100	108
9	102	NM	99	-	100	115
10	160	NM	94	-	100	111
11	25.7	NM	138	-	100	115
12	81.8	38	102	115	100	110
13	39.3	40	112	112	100	109
14	38.3	44.2	150	139	100	132
15	81.9	80.9	105	106	100	122
16	52.8	66.3	115	109	100	117
17	89.3	93.1	102	101	100	115
Average	88.4	68.2	107	110	100	114

NM-Not measured due to incorporation immediately after application by equipment located on the sprayer.

**Table VI. Half-lives calculated using zero-time values of 100%
and 50% of applied**

Study	Half-Life (days) with the Indicated Zero-Time Value*			
	100 % of Applied		50 % of Applied	
1	56.2	(38.6-103)	62.4	(40.3-138)
2	54.4	(42.4-75.8)	63.5	(49.5-88.4)
3	70.5	(55.9-95.4)	77.8	(58.0-118)
4	60.4	(42.3-106)	65.5	(44.4-125)
5	9.3	(7.2-13.2)	10.1	(7.7-14.6)
6	0.9	(0.6-2.1)	1.1	(0.5-6.1)
7	4.8	(2.1-14.7)	5.5	(2.6-60.8)
8	18.2	(12.0-37.8)	19.6	(13.3-37.4)
9	6.8	(4.5-13.4)	7.8	(5.6-12.5)
10	29.7	(18.1-82.8)	33.1	(21.7-69.7
11	5.2	(2.6-220)	6.0	(3.6-18.6)
12	9.3	(7.2-13.0)	10.2	(6.7-21.7)
13	38.2	(30.4-51.4)	41.5	(39.5-43.8)
14	28.7	(10.7-41.9)	37.8	(16.4-122)
15	36.0	(15.3-101)	44.0	(19.7-187)
16	139	(84-403)	163	(109-326)
17	94.6	(59.9-225)	109	(67.0-299)

*Numbers in parentheses provide the 90 % confidence interval

Conclusions

The zero-time core sample problem does not come with a simple solution since many processes contribute to low recoveries. Several different types of capture systems have shown that regardless of formulation or application method, the chemical is getting to the soil. There is no universal procedure that will guarantee good zero-time core samples. Later core samples are more representative because chemical has moved into the soil and is less prone to surface effect losses.

Low zero-time soil core residues result in only slightly longer, more conservative half-lives. It is necessary to maintain flexibility with regard to models used to analyze data, so that the result actual reflects the data, not an assumed degradation mechanism. Scientists continue in their efforts to improve core sampling but soil studies should not be invalidated because of low soil core results if proper application can be demonstrated by deposition samples.

Acknowledgments - ACPA Representatives in 1996

Tom Bregger	Agrevo	Mark Allan	Monsanto
Jim Devine	Am Cy	Monte Marshall	Monsanto
James Clark	BASF	Frank Norris	Rhône-Poulenc
Scott Jackson	BASF	Russell Jones	Rhône-Poulenc
Val Clay	Bayer	Berni Chong	Rohm & Hass
Bob Speth	Ciba	Charlie Keefter	Sandoz
Tom Wiepke	Ciba	Brent Solomon	Valent
Al Barefoot	Dupont	Diana Graham	Zeneca
Ron Cook	FMC	Paul Hendley	Zeneca

References

1. EPA Subdivision N Rejection Rate Analysis Meetings, 1992.
2. *Pesticide Reregistration Rejection Rate Analysis*, United States Environmental Protection Agency, EPA 738-R-93-010, 1993.

Appendices

Appendix 1. Averaged, adjusted* zero-time recovery data.

Product / Formu- lation	Soil** (% OM)	Lb a.i./A	Soil Core Samples Area (sq. inches)	Recovery (%)	Deposition Samples Substrate	Area (sq. inches)	Recovery (%)
A / -	SiL (1.1)	18.00	12.57	60.0	-	-	-
A / DF	S (0.0)	1.60	3.14	85.0	Soil	4.0	71.0
A / DF	-	2.00	3.14	78.0	Soil	4.0	120.0
A / WP	L (0.6)	1.08	12.57	91.0	Soil	4.0	108.0
A / WP	SiL (0.6)	1.08	12.57	57.0	Soil	4.0	102.0
B / EC	LS (1.5)	1.80	3.14	53.0	Soil	157.5	93.1
B / EC	CL (2.5)	2.00	3.14	75.0	Soil	157.5	91.3
C / DF	SL (1.3)	0.60	3.98	62.0	-	-	-
C / DF	SL (1.3)	0.60	-	-	Soil	9.6	87.0
C / DF	SiL (1.0)	0.90	-	-	Soil	9.6	103.0
C / DF	SiL (1.0)	0.90	3.98	82.0	-	-	-

Product / Formu-lation	Soil** (% OM)	Lb a.i./A	Soil Core Samples		Deposition Samples		
			Area (sq. inches)	Recovery (%)	Substrate	Area (sq. inches)	Recovery (%)
C / DF	L (1.1)	0.99	3.98	83.0	Paper	62.0	75.0
C / DF	SiL (1.0)'	0.99	3.98	62.0	Paper	62.0	73.0
C / DF	LS (0.5)	1.00	3.98	65.0	Paper	62.0	86.0
C / DF	LS (0.9)	1.00	3.98	42.0	Paper	62.0	71.0
C / DS	-	0.11	3.98	-	Paper	62.0	96.0
C / DS	L (4.0)	0.11	3.98	-	Paper	62.0	50.0
C / DS	L (4.0)	0.11	3.98	-	Paper	62.0	17.0
C / DS	SiL (6.2)	0.11	3.98	-	Paper	62.0	63.0
C / DS	SiL (6.2)	0.11	3.98	-	Paper	62.0	64.0
C / DS	SL (2.6)	0.11	3.98	-	Paper	62.0	72.0
C / DS	SL (2.6)	0.11	3.98	-	Paper	62.0	64.0
C / DS	SL (3.0)	0.11	3.98	-	Paper	62.0	82.0
C / DS	SL (3.0)	0.11	3.98	-	Paper	62.0	82.0
C / DS	L (1.6)	0.55	3.98	56.0	Paper	62.0	90.0
C / DS	LS (0.4)	0.55	3.98	66.0	Paper	62.0	89.0
C / DS	SL (0.3)	0.55	3.98	37.0	Paper	62.0	86.0
C / DS	SL (1.0)'	0.55	3.98	89.0	Paper	62.0	87.0
C / EC	S (0.3)	0.55	3.98	44.0	Paper	62.0	123.0
C / EC	SL (1.2)	0.99	3.98	27.0	Paper	62.0	96.0
C / EC	SL (1.2)	0.99	3.98	38.0	Paper	62.0	89.0
C / EC	SL (1.2)	0.99	3.98	66.0	Paper	62.0	93.0
C / EC	SL (1.2)	0.99	3.98	42.0	Paper	62.0	99.0
C / EC	SL (1.2)	0.99	3.98	50.0	Paper	62.0	82.0
C / EC	SL (1.2)	0.99	3.98	37.0	Paper	62.0	71.0
C / EC	SL (1.2)	0.99	3.98	32.0	Paper	62.0	79.0
C / EC	SL (1.2)	0.99	3.98	36.0	Paper	62.0	85.0
C / EC	SL (0.7)	1.28	3.98	68.0	Paper	62.0	63.0
C / EC	SL (0.7)	1.28	3.98	63.0	Paper	62.0	76.0
C / EC	SL (0.7)	1.28	3.98	53.0	Paper	62.0	78.0
C / EC	SL (0.7)	1.28	3.98	73.0	Paper	62.0	61.0
C / EC	SL (0.7)	1.28	3.98	118.0	Paper	62.0	61.0
C / EC	SL (0.7)	1.28	3.98	56.0	Paper	62.0	79.0
C / EC	SL (0.7)	1.28	3.98	60.0	Paper	62.0	87.0
C / EC	SL (0.7)	1.28	3.98	57.0	Paper	62.0	79.0
C / EC	S (0.9)	1.35	3.98	72.0	Paper	62.0	82.0
C / EC	S (0.9)	1.35	3.98	51.0	Paper	62.0	69.0
C / EC	S (0.9)	1.35	3.98	48.0	Paper	62.0	84.0
C / EC	S (0.9)	1.35	3.98	76.0	Paper	62.0	46.0
C / EC	S (0.8)	1.67	3.98	46.0	Paper	62.0	81.0
C / EC	S (0.8)	1.67	3.98	34.0	Paper	62.0	53.0
C / EC	S (0.8)	1.67	3.98	51.0	Paper	62.0	42.0
C / EC	S (0.8)	1.67	3.98	81.0	Paper	62.0	72.0

Product / Formu-lation	Soil** (% OM)	Lb a.i./A	Soil Core Samples		Deposition Samples		
			Area (sq. inches)	Recovery (%)	Substrate	Area (sq. inches)	Recovery (%)
C / EC	SL (0.9)	2.25	3.98	100.0	Paper	62.0	72.0
C / EC	SL (0.9)	2.25	3.98	100.0	Paper	62.0	72.0
C / EC	S (0.7)	2.27	3.98	38.0	Paper	62.0	-
C / EC	S (0.7)	2.27	3.98	43.0	Paper	62.0	-
C / EC	S (0.7)	2.27	3.98	67.0	Paper	62.0	96.0
C / EC	S (1.5)	11.00	-	-	Soil	9.6	78.0
C / EC	S (1.5)	11.00	3.98	97.0	-	-	-
C / EC	S (1.5)	11.00	-	-	Soil	169.1	73.0
C / G	S (1.4)	0.17	3.98	41.0	-	-	-
C / G	SL (4.9)	0.17	3.98	91.0	-	-	-
C / WP	SiL (1.0)	0.50	-	-	Soil	9.6	79.0
C / WP	SiL (1.0)	0.50	3.98	57.0	Paper	62.0	71.0
C / WP	SiL (1.0)	0.50	-	-	Soil	169.1	87.0
C / WP	SiL (2.5)	0.50	-	-	Soil	9.6	72.0
C / WP	SiL (2.5)	0.50	-	-	Soil	169.1	85.0
C / WP	SiL (2.5)	0.50	3.98	72.0	Paper	62.0	81.0
D / DF	CL (6.3)	0.03	-	85.3	Soil	9.6	87.2
E / -	-	0.50	-	52.9	Soil	7.1	72.9
E / -	-	0.51	-	54.1	Soil	7.1	62.2
E / -	-	0.51	-	83.1	Filter Paper	23.8	74.2
E / -	-	0.51	-	143.9	Filter Paper	23.8	110.5
F / -	L (2.4)	-	4.15	69.0	Petri Dish	9.6	68.0
F / -	L (3.6)	-	4.15	100.0	Petri Dish	9.6	108.0
F / -	L (3.9)	-	4.15	131.0	Petri Dish	9.6	84.0
F / -	L (5.2)	-	4.15	96.0	Petri Dish	9.6	48.0
F / -	SiL (3.0)	-	4.15	100.0	Petri Dish	9.6	70.0
F / -	SL (4.1)	-	4.15	81.0	Petri Dish	9.6	26.0
H / -	-	-	-	-	Filter Paper	-	85.0
H / -	-	-	-	-	Filter Paper	-	89.3
H / -	-	-	-	-	Soil	-	163.2
H / -	-	-	-	-	Soil	-	67.4
H / -	-	0.04	-	-	Petri Dish	9.6	76.0
H / -	-	0.13	-	-	Petri Dish	9.6	92.0
H / -	-	0.13	-	-	Petri Dish	9.6	86.0
H / -	-	0.13	-	-	Petri Dish	9.6	93.0
H / -	-	0.13	-	-	Petri Dish	9.6	80.0
H / -	-	0.21	-	-	Petri Dish	9.6	83.0
H / -	-	0.21	-	-	Petri Dish	9.6	83.0
H / -	-	0.21	-	-	Petri Dish	9.6	40.0
H / -	-	0.21	-	-	Petri Dish	9.6	70.0
H / -	-	0.21	-	-	Petri Dish	9.6	68.0
H / -	-	0.21	-	-	Petri Dish	9.6	38.0

Product / Formu-lation	Soil** (% OM)	Lb a.i./A	Soil Core Samples		Deposition Samples		
			Area (sq. inches)	Recovery (%)	Substrate	Area (sq. inches)	Recovery (%)
H / -	-	0.25	-	-	Soil	-	36.0
H / -	-	0.25	-	-	Filter Paper	-	80.9
H / -	-	0.25	-	-	Filter Paper	-	66.3
H / -	-	0.25	-	-	Soil	-	77.1
H / -	-	0.25	-	-	Filter Paper	-	44.2
H / -	-	0.25	-	-	Filter Paper	-	93.1
H / -	-	0.25	-	-	Soil	-	50.6
H / -	-	0.25	-	-	Soil	-	84.0
H / -	-	0.29	-	-	Petri Dish	9.6	80.0
H / -	-	0.29	-	-	Petri Dish	9.6	78.0
H / -	-	0.29	-	-	Petri Dish	9.6	50.0
H / -	-	0.29	-	-	Petri Dish	9.6	95.0
H / -	-	0.67	-	-	Petri Dish	9.6	79.0
H / -	-	0.67	-	-	Petri Dish	9.6	84.0
H / -	-	0.67	-	-	Petri Dish	9.6	66.0
H / -	-	0.75	-	-	Soil	-	90.0
H / -	-	0.75	-	. -	Soil	-	138.4
H / -	-	0.75	-	-	Filter Paper	-	80.0
H / -	-	0.75	-	-	Filter Paper	-	79.2
H / -	-	0.75	-	-	Filter Paper	-	78.5
H / -	-	0.75	-	-	Soil	-	124.2
H / -	-	1.00	-	-	Filter Paper	-	85.3
H / -	-	1.00	-	-	Filter Paper	-	82.8
H / -	-	1.00	-	-	Filter Paper	-	94.6
H / -	-	1.00	-	-	Filter Paper	-	90.2
H / -	-	1.00	-	-	Filter Paper	-	81.1
H / -	-	1.00	-	-	Soil	-	87.2
H / -	-	1.00	-	-	Filter Paper	-	82.5
H / -	-	1.87	-	-	Petri Dish	9.6	25.0
H / -	-	2.00	-	-	Filter Paper	-	131.8
H / -	-	2.00	-	-	Filter Paper	-	81.6
H / -	-	2.00	-	-	Filter Paper	-	79.3
H / -	-	2.00	-	-	Filter Paper	-	88.3
H / -	-	2.00	-	-	Soil	-	83.2
H / -	-	2.00	-	-	Soil	-	101.0
H / -	-	2.00	-	-	Soil	-	76.8
H / -	-	2.00	-	-	Soil	-	80.0
H / -	-	2.14	-	-	Petri Dish	9.6	76.0
H / -	-	2.14	-	-	Petri Dish	9.6	103.0
H / -	-	2.14	-	-	Petri Dish	9.6	62.0
H / -	-	2.14	-	-	Petri Dish	9.6	106.0
H / -	-	2.23	-	-	Petri Dish	9.6	74.0

Product / Formu-lation	Soil** (% OM)	Lb a.i./A	Soil Core Samples		Deposition Samples		
			Area (sq. inches)	Recovery (%)	Substrate	Area (sq. inches)	Recovery (%)
J / WDG	-	-	-	68.6	-	-	-
J / WDG	-	-	-	149.2	Filter Paper	9.6	93.2
J / WDG	-	-	-	80.0	Filter Paper	9.6	86.9
J / WDG	-	-	-	71.4	Pan	120.0	92.6
J / WP	-	-	-	-	Filter Paper	9.6	81.5
J / WP	-	-	-	-	Filter Paper	9.6	69.4
J / WP	-	-	-	-	Filter Paper	9.6	83.8
J / WP	-	-	-	-	Filter Paper	9.6	102.2
J / WP	-	-	-	-	Pan	120.0	93.3
J / WP	-	-	-	-	Filter Paper	9.6	110.6
J / WP	-	-	-	-	Soil	9.6	77.4
J / WP	-	-	-	100.0	Soil	9.6	109.4
J / WP	-	-	-	-	Filter Paper	9.6	75.5
J / WP	-	-	-	-	Filter Paper	9.6	80.0
J / WP	-	-	-	-	Filter Paper	9.6	78.2
J / WP	-	-	-	-	Filter Paper	9.6	79.0
J / WP	-	-	-	76.5	Soil	9.6	102.2
J / WP	-	-	-	83.7	Filter Paper	9.6	75.0
K / -	-	-	-	51.0	Petri Dish	9.6	88.0
K / -	-	-	-	65.0	Petri Dish	23.7	82.0
L / -	-	0.20	-	-	Petri Dish	9.6	93.4
L / -	-	0.24	-	-	Petri Dish	9.6	128.3
L / -	-	0.24	-	-	Petri Dish	9.6	107.5
L / -	-	0.24	-	-	Petri Dish	9.6	110.3
L / -	-	0.25	-	-	Petri Dish	9.6	109.8
L / -	-	0.25	-	-	Petri Dish	9.6	82.3
L / -	-	0.25	-	-	Petri Dish	9.6	112.3
L / -	-	0.90	-	-	Petri Dish	9.6	98.0
L / -	-	0.90	-	-	Petri Dish	9.6	114.3
L / -	-	0.90	-	-	Petri Dish	9.6	101.0
M / -	C (1.7)	6.00	19.63	55.0	Poly Foam Pad	864.0	95.0
M / -	L (34.0)	6.00	19.63	35.0	Poly Foam Pad	864.0	98.0
M / -	LS (1.2)	6.00	19.63	55.0	Poly Foam Pad	864.0	85.0
M / -	L (2.5)	9.59	22.06	43.0	Poly Foam Pad	1296.0	93.0
M / -	L (3.2)	9.88	22.06	49.0	Poly Foam Pad	1296.0	89.0
M / -	SL (1.0)	9.99	22.06	63.0	Poly Foam Pad	1296.0	96.0
M / -	SL (3.1)	10.02	22.06	64.0	Poly Foam Pad	1296.0	96.0
M / EC	LS (1.7)	3.00	19.63	60.0	-	-	-
M / EC	LS (1.7)	3.00	3.14	28.0	Filter Paper	180.0	59.0
M / EC	CL (3.8)	3.40	19.63	65.0	-	-	-
M / EC	CL (3.8)	3.40	3.14	67.0	Filter Paper	180.0	102.0
M / EC	SL (2.2)	4.00	19.63	65.0	Poly Foam Pad	864.0	103.0

Product / Formu- lation	Soil** (% OM)	Lb a.i./A	Soil Core Samples		Deposition Samples		
			Area (sq. inches)	Recovery (%)	Substrate	Area (sq. inches)	Recovery (%)
M / G	L (3.4)	6.00	19.63	54.0	Poly Container	600.0	150.0
M / G	LS (1.2)	6.20	19.63	65.0	Poly Container	600.0	100.0
M / G	C (1.7)	6.30	19.63	74.0	Poly Container	600.0	94.0
M / ME	LS (1.7)	1.60	3.14	37.0	Filter Paper	180.0	61.0
M / ME	CL (3.8)	3.00	19.63	69.0	-	-	-
M / ME	CL (3.8)	3.00	3.14	36.0	Filter Paper	180.0	77.0
M / WDG	-	0.04	-	-	Poly Foam Pad	1296.0	86.0
M / WDG	-	0.04	-	-	Poly Foam Pad	1296.0	101.0
M / WP	SL (0.0)	0.32	25.97	22.0	-	-	-
M / WP	SL (0.0)	0.32	3.14	89.0	-	-	-
M / WP	S (0.4)	0.34	3.14	74.0	-	-	-
M / WP	S (0.4)	0.34	5.31	0.4	-	-	-
Average		1.7	7.0	65.3		128.8	83.3
Std Deviation		2.6	6.4	24.9		291.6	20.6
Maximum Value		18.0	25.97	149.2		1296.0	163.2
Minimum Value		0.03	3.14	0.4		4.0	17.0
Number		161	89	94		144.0	175

* Data provided by various companies were standardized to facilitate comparison using the following criteria:

1. Where ranges were given, the mid-point of the range was used.

2. Where average +/- std. dev. was given, the average was used.

3. Diameter expressed as "NACA Std." was assumed to have a diameter of 2.3 inches.

4. Unspecified petri dish diameter assumed to be 89 mm.

5. Data from the same studies have been averaged by sampling device.

** Soil codes: C=Clay, CL=Clay Loam, L=Loam, LS=Loamy Sand, S=Sand, SL=Sandy Loam, SiL=Silt Loam (see Appendix 2).

Appendix 2. Soil characteristics

Soil Code	Soil Type (0-6 inch)	% Sand	% Silt	% Clay	% Organic Matter (OM)
C (1.7)	Clay	35.0	23.0	42.0	1.7
CL (2.5)	Clay Loam	30.0	42.0	31.0	2.5
CL (3.8)	Clay Loam	22.0	44.0	34.0	3.8

Soil Code	Soil Type (0-6 inch)	% Sand	% Silt	% Clay	% Organic Matter (OM)
CL (6.3)	Clay Loam	36.0	30.0	34.0	6.3
L (0.6)	Loam	51.2	38.0	10.8	0.6
L (1.1)	Loam	49.2	32.4	18.4	1.1
L (1.6)	Loam	47.2	33.2	19.6	1.6
L (2.4)	Loam	44.0	44.0	12.0	2.4
L (2.5)	Loam	40.0	43.0	17.0	2.5
L (34.0)	Loam	32.0	49.0	19.0	34.0
L (3.2)	Loam	36.0	37.0	27.0	3.2
L (3.4)	Loam	32.0	49.0	19.0	3.4
L (3.6)	Loam	44.0	40.0	16.0	3.6
L (3.9)	Loam	35.0	45.0	20.0	3.9
L (4.0)	Loam	33.2	45.6	21.2	4.0
L (5.2)	Loam	49.0	36.0	15.0	5.2
LS (0.4)	Loamy Sand	79.2	11.2	9.6	0.4
LS (0.5)	Loamy Sand	88.0	4.0	8.0	0.5
LS (0.9)	Loamy Sand	78.4	14.8	6.8	0.9
LS (1.2)	Loamy Sand	84.0	9.0	7.0	1.2
LS (1.5)	Loamy Sand	81.5	11.5	7.5	1.5
LS (1.7)	Loamy Sand	86.0	12.0	2.0	1.7
S (0.0)	Sand	-	-	-	-
S (0.3)	Sand	90.8	3.2	6.0	0.3
S (0.4)	Sand	90.0	8.0	2.0	0.4
S (0.7)	Sand	96.7	2.0	1.3	0.7
S (0.8)	Sand	94.7	4.0	1.3	0.8
S (0.9)	Sand	90.7	8.0	1.3	0.9
S (1.4)	Sand	88.0	8.0	4.0	1.4
S (1.5)	Sand	98.0	1.0	1.0	1.5
SiL (0.6)	Silt Loam	16.8	66.4	16.8	0.6
SiL (1.0)	Silt Loam	27.0	19.0	19.0	1.0
SiL (1.0)'	Silt Loam	22.8	52.4	24.8	1.0
SiL (1.1)	Silt Loam	35.2	51.6	13.2	1.1
SiL (2.5)	Silt Loam	24.0	25.0	25.0	2.5

Soil Code	Soil Type (0-6 inch)	% Sand	% Silt	% Clay	% Organic Matter (OM)
SiL (3.0)	Silt Loam	39.0	51.0	10.0	3.0
SiL (6.2)	Silt Loam	29.2	53.6	17.2	6.2
SL (0.0)	Sandy Loam	-	-	-	-
SL (0.3)	Sandy Loam	76.8	9.2	14.0	0.3
SL (0.7)	Sandy Loam	65.2	23.8	11.2	0.7
SL (0.9)	Sandy Loam	77.2	14.4	8.4	0.9
SL (1.0)	Sandy Loam	56.0	34.0	10.0	1.0
SL (1.0)'	Sandy Loam	79.2	11.2	9.6	1.0
SL (1.2)	Sandy Loam	71.2	17.2	11.2	1.2
SL (1.3)	Sandy Loam	73.0	12.0	12.0	1.3
SL (2.2)	Sandy Loam	64.0	28.0	8.0	2.2
SL (2.6)	Sandy Loam	54.8	28.4	16.8	2.6
SL (3.0)	Sandy Loam	55.2	28.4	16.4	3.0
SL (3.1)	Sandy Loam	62.0	24.0	14.0	3.1
SL (4.1)	Sandy Loam	56.0	32.0	12.0	4.1
SL (4.9)	Sandy Loam	69.0	22.0	9.0	4.9

Chapter 6

Sources and Magnitudes of Variability in the Terrestrial Field Dissipation of Pesticides

Joseph H. Massey[1] and James S. LeNoir[2]

[1]Department of Plant and Soil Sciences, Mississippi State University, Mississippi State, MS 39762
[2]Stine-Haskell Research Center, DuPont Agricultural Products, DuPont, Newark, DE 19714

Excessive within-site variability in terrestrial field dissipation rate data can prevent the accurate determination of a pesticide's rate of dissipation. The primary sources of variability in these studies include errors associated with test substance application ($s^2_{Application}$), soil sampling ($s^2_{Sampling}$), sample analysis ($s^2_{Analysis}$) and variability among factors affecting abiotic and biotic dissipation processes in soil ($s^2_{Dissipation}$). The physicochemical properties of a pesticide also affect the observed variability in dissipation rate. Of these potential sources of error, $s^2_{Application}$ and $s^2_{Analysis}$ are the best characterized, each yielding CV values of 10 to 15% for well-controlled bare-soil studies. Variability induced during soil sample homogenization may represent a significant portion of $s^2_{Analysis}$. The magnitudes of $s^2_{Sampling}$ and $s^2_{Dissipation}$ are not well-characterized; both may represent significant sources of variability under field conditions. In practice, it may be difficult to differentiate $s^2_{Sampling}$ from $s^2_{Dissipation}$ due to confounding, but one possible approach is offered here.

Introduction

Excessive within-site variability in terrestrial field dissipation (TFD) studies can prevent the accurate determination of a pesticide's dissipation rate, and has historically ranked seventh out of fifteen rejection factors for TFD studies submitted to the U.S. Environmental Protection Agency (1). Study rejection can result in expensive repeat studies and jeopardized registration timelines.

The amounts of pesticide measured in soil at a given time vary spatially with distance and depth, resulting in variation within and between treated replicates. Data exhibiting variability, such as shown in Figure 1, provoke these questions: "What is the primary source(s) of this variability? Is the majority of this variability controllable or does it arise from 'natural' sources over which we have little practical control?" "What, if anything, might we do different in the future to avoid such variability?" The objectives of this exercise were to (a) evaluate the sources and relative magnitudes of this variability and, (b) discuss techniques for differentiating "induced" and "inherent" sources of variability in TFD studies. A better understanding of this variability could lead to improved study design and, ultimately, less variable dissipation rate data.

Defining Total Variability(s^2_{Total})

In order to answer the above questions, one must first define "variability" as it pertains to TFD studies and determine how it will be measured in this particular context. Researchers typically arrange their experimental units so as to minimize within-block (i.e., within-replicate) variation while maximizing between-block variation. From a TFD study diagnostic perspective, it therefore seems reasonable to define total variability (s^2_{Total}) as the variance in pesticide residue levels existing between uncomposited soil cores collected within the same replicate at a given sampling period, i.e.,

$s^2_{Total} \equiv$ the within-replicate variance existing between pesticide residue levels measured in uncomposited soil cores.

This within-replicate variability occurs as a result of errors associated with test substance application ($s^2_{Application}$), soil sampling ($s^2_{Sampling}$), sample analysis ($s^2_{Analysis}$) as well as spatial variability among factors affecting the abiotic and biotic transformation, and physical redistribution, processes in soil ($s^2_{Dissipation}$). A generic model for this total variability is:

$$s^2_{Total} = f\left(s^2_{Application}, s^2_{Sampling}, s^2_{Analysis}, s^2_{Dissipation}\right) \qquad (1)$$

where s^2_{Total}, $s^2_{Application}$, $s^2_{Sampling}$, $s^2_{Analysis}$, and $s^2_{Dissipation}$ are variances having the common units of mass/area (e.g., (g a.i./ha)^2). It is important that all residues are normalized according to the sampled area so that the differences in the concentrations existing between soil cores and application monitors, for example, do not affect calculations involving the different sources of variation.

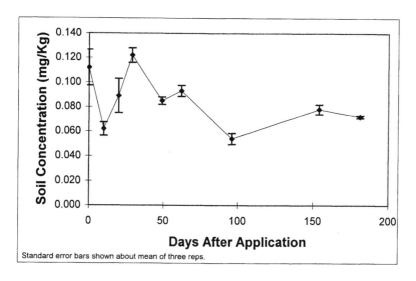

Figure 1. Example of variable terrestrial field dissipation data set

The exact relationships between these sources of variability are not known. However, if the sources of variability are assumed to be additive in nature, then Equation 1 becomes

$$s^2_{Total} = s^2_{Application} + s^2_{Sampling} + s^2_{Analysis} + s^2_{Dissipation}. \qquad (2)$$

In the event that s^2_{Total} is much greater than the summed variances given in Equation 2, then this might suggest that there is a source(s) of variability missing from the model. In the event that s^2_{Total} is much less than the summed variances, this might suggest that the relationship between the different sources of variability is non-additive.

Placing the previous questions in more quantitative terms, we want to know what proportion of the within-replicate variability existing between pesticide levels measured in soil may be attributed to error sources that are controllable (e.g., "induced") versus that which is due solely to natural soil variations occurring across the field (e.g., "inherent" variability). Thus the proportion of "induced" variability is given by

$$(s^2_{Application} + s^2_{Sampling} + s^2_{Analysis})/(s^2_{Total}). \qquad (3)$$

The proportion of "inherent" variability is given by

$$(s^2_{Dissipation})/(s^2_{Total}). \qquad (4)$$

Materials and Methods

Estimating Application Error ($s^2_{Application}$)

The accuracy and precision of test substance applications is commonly measured using application monitors consisting of either empty or soil-filled containers, or disks made of foam or paper. The main benefit of application monitors is that they allow for the measurement of spray deposition without the errors associated with the collection and processing of soil cores. For the purpose of study diagnostics, application error is determined by the analysis of multiple, uncomposited application monitors placed within the treated soil replicates (i.e., $s^2_{Application} \approx$ variance between individually-analyzed application monitors). The use of application monitors to determine application error has become routine in the conduct of regulatory TFD studies.

Estimating Analytical Error ($s^2_{Analysis}$)

Errors arising in sample analysis can occur in any of three main phases: (a) sample processing (includes soil core weight determinations, compositing and blending; moisture determinations, aliquot subsampling and weighing), (b) sample extraction, and (c) sample analysis. The error associated with these phases can be estimated by determining the variance existing between duplicate analyses (i.e., $s^2_{Analysis} \approx$ variance between duplicate soil analyses). Duplicate sample analyses are routinely employed in standard analytical protocols for TFD studies.

Estimating Soil Sampling Error ($s^2_{Sampling}$)

In order to estimate the contribution that sampling error makes to s^2_{Total}, one must assume that dissipation of the compound at zero-time did not occur (i.e., $s^2_{Dissipation} \approx 0$ at zero-time). This is a fair assumption for relatively stable, non-volatile compounds and can be confirmed by the absence of degradation products in the zero-time soil cores. If this assumption holds, then Equation 2 becomes:

$$s^2_{Total} = s^2_{Application} + s^2_{Sampling} + s^2_{Analysis} \qquad (5)$$

Because s^2_{Total} represents the variance existing between residue levels in within-replicate samples, and $s^2_{Application}$ and $s^2_{Analysis}$ can be determined as above, $s^2_{Sampling}$ can be estimated by subtraction at zero-time as follows

$$s^2_{Sampling} = s^2_{Total} - s^2_{Application} - s^2_{Analysis.} \qquad (6)$$

This approach was used by Holwarth et al. (2) to estimate soil sampling error and requires the analysis of uncomposited, zero-time soil cores and the normalization of all residues on a mass/area basis.

Estimating the Variability of Soil Dissipation Processes ($s^2_{Dissipation}$)

To estimate $s^2_{Dissipation}$ using an approach similar to that for $s^2_{Sampling}$, it must be assumed that $s^2_{Sampling}$ is constant over time. In other words, one assumes that the same soil sampling crew and sampling equipment are used and that soil conditions are similar at each sampling period. Clearly, the validity of this assumption needs additional evaluation and may not hold true. However if it does, estimates of $s^2_{Dissipation}$ could be made using non-composited cores collected at a time periods > zero-time as follows

$$[s^2_{Dissipation}]_{t\,>\,0} = [s^2_{Total}]_{t\,>\,0} - [s^2_{Application}]_{t=0} - [s^2_{Sampling}]_{t=0} - [s^2_{Analysis}]_{t>0} \qquad (7)$$

where $[s^2_{Dissipation}]_{t\,>\,0}$ is the estimated variance due to soil dissipation processes occurring at a time period > zero-time, $[s^2_{Total}]_{t\,>\,0}$ is the total variance between residue levels at this time, $[s^2_{Application}]_{t=0}$ is the application error as previously determined, $[s^2_{Sampling}]_{t=0}$ is the sampling error determined at zero-time, and $[s^2_{Analysis}]_{t>0}$ is the analytical error determined from duplicate analyses measured at the same time period as for the total and dissipation variance values.

Primary Data Source

In 1997, seven bare-soil TFD studies for a developmental compound formulated as emulsifiable concentrate were conducted at four U.S. sites and three European sites. The U.S. field studies were conducted by different research crews. The same application and sampling crew conducted all three of the European studies. Test substance applications were made to three treated replicates at a rate of 495 g a.i./ha using an application volume of ca. 187 L/ha and flat-fan nozzles. Five soil cores (5.1 cm diameter) per untreated control and treated replicate were collected per sampling period. The soil samples were extracted using accelerated solvent extraction and cleanup was performed using solid-phase extraction. The residues were quantified by GC-MS. Data utilized from these studies consisted of pesticide concentrations (mg/Kg) in soil cores (0- to 15-cm only) collected over a period of ca. 180 days after treatment. In addition, nine 10 x 10-cm α-cellulose (Schleicher and Schuell, Inc., Keene, NH) application monitors were used at each study location.

Results and Discussion

Influence of Pesticide Physicochemical Properties on Dissipation Variability

The physicochemical properties of a pesticide largely determines how it will behave in a given environment. As a result, the dissipation of one compound may be inherently more variable than another simply due to differences in physicochemical properties. When Mojasevic and Helling (3) investigated the variability in zero-time soil recoveries of six simultaneously-applied pesticides, for example, they found that metribuzin and cyanazine both exhibited higher overall variability than alachlor, atrazine, ethoprophos, and metolachlor whose coefficient of variation (CV) values ranged from 12 to 35%. The investigators found that differences in physicochemical properties existing between the compounds affected the overall variabilities observed in this bare-soil field dissipation study. Pesticides that are volatile[1] or otherwise chemically and/or biologically unstable present unique challenges in terms of the accurate determination of field soil dissipation rates. Clearly, the specific chemical and physical properties of the test substance(s) must always be taken into consideration during the design and conduct of terrestrial field dissipation studies.

All aspects of field study conduct, from test substance application to sample analysis, have the potential to induce significant amounts of variability when not carefully performed. Below are estimates of key sources of variability affecting the bare-soil terrestrial field dissipation of pesticides:

Test-Substance Application Error ($s^2_{Application}$)

The variability associated with test substance application is one of the best-characterized sources of variability in TFD studies. Potential sources of variability include clogged or worn spray nozzle(s), poor pressure regulation, non-homogeneous spray solution, overlapping or missed areas in multiple boom-

---

[1] Volatilization from soil is an inherently variable process that is strongly affected by variations in soil organic matter and moisture contents, temperature and wind speed (4).

width applications, and drift due to variable wind gusts[2]. Spray nozzle type impacts spray droplet size distribution and, hence, the potential for drift. In their analysis of over 1,600 pesticide applications made for a variety of GLP crop residue trials, Braverman et al. (5) found that the factors most responsible for *inaccurate* pesticide applications (i.e., application rates applied >10% or <5% of the target rate) were improper boom height (60% of errors), miscalculation of application rate (26% of errors), and variation in pass time (14% of errors). In terms of application *precision*, variations in boom height and travel speed during application can also induce significant overall variability in soil residue levels measured in TFD studies.

The CV for new spray nozzles is typically ≤ 7% under laboratory conditions (6). This represents the lower-limit of variability that should be expected for applications performed under field conditions. In the present study, the CV values between nine individually analyzed α-cellulose spray monitors for the sites averaged 16% and ranged from 11 to 29% (Table 1). These values resulted from five different applicators using three types of application equipment. Our results agreed with reported CV values that ranged from 12% for polyurethane foam monitors to 15% for soil-filled petri dishes used in a large number of TFD studies (7). Based on these observations, application error under conditions typical of TFD studies yields CV values on the order of 15% to 20%[3]. Application variability significantly greater than this warrants further investigation as to its cause.

For agrochemicals requiring soil incorporation after application, special care must be taken to ensure that the residues are evenly and thoroughly incorporated. Researchers have observed CV values of 50 to 80% for residues after incorporation by discing (8, 9). This variability was largely attributed to incomplete incorporation after application. Regardless of the type of equipment used, three incorporation passes are typically required for thorough incorporation of soil-applied pesticides (10).

---

[2] Spray drift has been found to vary from 1.4% of the applied spray volume under "ideal" spray conditions to >37% under "less-than-ideal" conditions (11).
[3] While the errors associated with granular formulations are not addressed in this exercise, evidence suggests that zero-time recoveries of granular formulations from soil can be highly variable (CV ≥80%) (3).

Sample Analysis Error($s^2_{Analysis}$)

Like application error, analytical error is among the best-characterized sources of variability in TFD studies. Potential sources of analytical variability are associated with sample storage, processing and subsampling, extraction, cleanup and instrumental analysis. Increasing subsample size generally reduces subsampling errors and provides subsamples more representative of the larger soil sample (12). Extraction efficiency can be impacted by variations in soil moisture, texture and organic matter content, among other factors. In the present study, $s^2_{Analysis}$ yielded a CV value of 11.5% averaged across six analytes extracted from seven soils (data not shown). Similarly, Trubey et al. (13) used standard extraction techniques and LC/MS/MS to determine the amounts of sulfometuron-methyl and five degradates in two soils, resulting in CV values of 11.4%. In general, analysis errors occurring within well-controlled soil methods typical of TFD studies yield CV values on the order of 10 to 15%.

In a separate laboratory study involving six soils (including the European soils from the present study), six 1-Kg amounts of each soil were individually fortified with 0.100 mL of a ^{14}C-labeled pesticide solution. After blending the soils for 20 min in an institutional-grade food mixer, seven 1-g aliquots were collected from each blended soil and combusted. Based on these combustion analyses, soil-blending error ranged from 8 to 18% CV (data not shown). These results suggest that significant error may arise during the blending of soil samples and support efforts aimed at ensuring the thorough blending of soil prior to sub-sampling and analysis.

Soil Sampling Error($s^2_{Sampling}$)

It has long been recognized that variability due to soil sampling is generally much greater than that associated with sample analysis (14). Soil sampling error is one of the least characterized sources of variability in TFD studies. Soil texture, soil-surface condition (e.g., surface roughness, soil moisture content), and sampling technique influence the magnitude of soil-sampling error. Potential sources of soil sampling variability include rough soil surfaces (e.g., stones, soil clods and/or crop debris) that interfere with representative sampling of surface residues, changes in soil moisture, and impenetrable objects or layers in the soil profile resulting in less-than-full soil cores. Assuming proper soil surface preparation, core diameter appears to have negligible impact on observed residue variability or average residue concentration (2, 8, 15).

In the present study, five zero-time samples were blended together to yield one composite sample for each of three replications. Therefore, an estimate of sampling error using the approach outlined in Equation 6 was not possible. Holzwarth et al. (2), however, used this approach to estimate sampling error for a silty loam soil and found $s^2_{Sampling}$ to be the single largest contributor to the total variability existing between zero-time soil residues for three different sampling probes (Table 2). In their study, $s^2_{Sampling}$ accounted for 42 to 83% of the total variance in ten individually analyzed soil cores. Interestingly, soil moisture had little impact on sampling error.

An alternative definition of soil sampling error was suggested by the work of Walker and Brown (15). These authors suggest that sampling error is related to the variance existing between the different weights of soil cores. In their study, core weights varied by less than 5%, suggesting that $s^2_{Sampling}$ was not significant in terms of total variation. Estimating $s^2_{Sampling}$ using variations in core weights, however, assumes the even distribution of pesticide residues throughout the entire soil core. This is not likely under most field study conditions. The approach outlined in Equation 6 is likely a more defensible approach for estimating sampling error.

Variability in Abiotic and Biotic Dissipation Processes ($s^2_{Dissipation}$)

There has been considerable interest in the spatial variability of pesticide residues and various pesticide dissipation processes in field soils (15-20). The impact of this variability on results specifically obtained by TFD studies, however, has not been well characterized. A number of soil properties and environmental conditions affect the dissipation rates of a pesticide in soil: Rates of microbial degradation, leaching, hydrolysis, photolysis and volatility are all influenced by soil texture, pH, organic matter content, moisture and oxygen status, and soil temperature. Of these soil parameters, bulk density, temperature and pH are generally among the least variable (CV <15%) while soil texture, organic matter and fertility are often moderately variable (Table 3). Variability in pesticide dissipation has been attributed primarily to differences in soil moisture and temperature (27). The most variable soil properties (CV>50%), however, are those related to water and solute movement (22); variability is also very high for microbial processes occurring within soil micro-sites (23, 24).

The variability of soil tends to increase with depth. Lafrance and Banton (28) found that the CV for organic carbon (OC) content was lower in the plow zone than for OC contents deeper in the soil profile. Moreover, they suggest that while OC is often considered to control pesticide sorption and leaching,

Table 1. Application variations at seven terrestrial dissipation field sites.

Site No.	Application Boom Type	Application Monitor CV (%)[1]
1	ATV-mounted	10.9
2	2-man hand-held	12.2
3	Tractor-mounted	16.2
4	Tractor-mounted	28.8
5	Bicycle wheel-	11.2
6	mounted	13.3
7		18.7

[1]Coefficient of variation (CV) for nine (9) individually analyzed spray monitors per application per site. Each spray monitor consisted of a 10 x 10-cm square of α-cellulose paper.

Table 2. Effect of soil sampling technique and soil moisture on zero-time concentrations of isoproturon in a silt-loam soil.

Soil Corer Type	Soil Moisture	Total Variance[1]	Application Variance[2]	Analytical Variance[3]	Sampling Variance[4]
2.9-cm	dry	395.9	81.5	26.9	287.6
diameter	wet	320.5	81.5	105.2	133.8
4.8-cm	dry	210.6	81.5	32.7	96.4
diameter	wet	626.0	81.5	23.3	521.3
6.6-cm bulb	dry	246.2	81.5	60.4	104.3
planter	wet	387.6	81.5	46.8	259.4

[1] Total variance (s^2) between 10 individually analyzed soil cores per corer type and moisture status. All variance values have units of $(mg\ a.i.\ /m^2)^2$.

[2] Variance between five individually analyzed application monitors.

[3] Analytical variance between duplicate samples.

[4] Variance due to sampling calculated by subtraction (see Equation 6).

variability in hydraulic conductivity likely plays a more important role in inducing variability in the physical redistribution of pesticides than variability in sorption parameters. Because large variations in the hydrologic properties of field soils often exist on very small spatial scales, it may be impossible to separate sampling errors from inherent dissipation variability. This may be especially true for compounds susceptible to redistribution and transport within the soil profile.

In the present study, we were unable to estimate $s^2_{Sampling}$ and, thus, were unable to determine if the approach outlined in Equation 7 would lead to useful estimates of $s^2_{Dissipation}$. Therefore, another approach was used to estimate the magnitude of within-site variability due to the inherent variability in soil dissipation processes. The CV values of first-order rate constants (k) were calculated using three treated replicates per site (Table 4). The CV values ranged from 2 to 62%. These results generally agree with similar values calculated for metribuzin (21%) and simazine (7%) (15), sulfometuron-methyl (41%) (13) and carbofuran (14%) (24). Relative to the other sources, $s^2_{Dissipation}$ clearly has the potential to represent a significant source of variability in TFD studies. The magnitude of $s^2_{Dissipation}$ will depend upon the level of variability existing in the field for the dissipation mode(s) relevant to the pesticide of interest. The impact of excessive variability on dissipation rate estimation is most detrimental for compounds undergoing relatively slow dissipation in soil (8). Unfortunately, there seems to be few techniques that allow the spatial variability of soils to be addressed in a manner practical to the conduct of regulatory TFD studies. Research suggests that reducing within-replicate variability below 20 to 25% is not likely using reasonable sampling rates (8, 15).

Summary & Conclusions

The primary sources of within-replicate variability in terrestrial field dissipation studies include errors associated with test substance application ($s^2_{Application}$), soil sampling ($s^2_{Sampling}$), sample analysis ($s^2_{Analysis}$) and variability among factors affecting abiotic and biotic dissipation processes in soil ($s^2_{Dissipation}$). Of these, $s^2_{Application}$ and $s^2_{Analysis}$ are the best defined, each yielding CV values of 10 to 15% CV for well-controlled studies. Special care must be taken during the incorporation of soil-applied pesticides to avoid the introduction of significant residue variability. Similarly, variability associated with incomplete soil homogenization in the laboratory likely represents a significant portion of $s^2_{Analysis}$. The magnitudes of $s^2_{Sampling}$ and $s^2_{Dissipation}$ are not well characterized but may represent significant sources of variability under field

Table 3. Within-field variabilities of soil properties affecting pesticide dissipation.

Soil Property	Coefficient of Variation (CV)
Bulk density; porosity; saturated water content; temperature; pH	Low variation: CV < 15%
Sand; silt; clay; organic matter;cation exchange capacity; water retention at various tensions; levels of nitrogen, phosphorus and potassium; Freundlich K	Mid-variation: 15% < CV < 50%
Saturated and unsaturated hydraulic conductivity; pore water velocity; electrical conductivity; denitrification and related microbial enzyme activities	High variation: CV > 50%

Notes: Table 3 was adapted from Reference 21. Additional data were compiled from References 18, and 22 through 26. Actual hydrogen ion activities in soil solution vary significantly more than indicated by CV values calculated using pH values.

Table 4. Between-replicate variation between first-order rate constants for seven terrestrial field dissipation studies.

Site No.	Soil Texture	First-Order Rate Constants (day^{-1})				
		Rep 1	Rep 2	Rep 3	Average	CV(%)[1]
1	Loamy sand	0.002	0.003	0.003	0.003	11.5
2	Silt loam	0.001	0.005	0.005	0.004	62.0
3	Silt loam	0.031	0.03	0.030	0.031	1.6
4	Sand	0.004	0.002	0.002	0.003	52.5
5	Sandy loam	0.010	0.009	0.007	0.009	14.7
6	Silt loam	0.004	0.003	0.003	0.004	12.6
7	Clay loam	0.001	0.001	0.002	0.001	41.3

[1]Represents the coefficient of variation (CV) between the first-order rate constants calculated for each individual replicate.

conditions. In practice, however, it may be difficult to differentiate $s^2_{Sampling}$ from $s^2_{Dissipation}$ due to the potentially confounding natures of these errors. Because few TFD studies are designed to allow a more thorough investigation into these matters, it is recommended that at least one replicate of soil cores collected at zero-time, and at a later sampling time(s), remain uncomposited. A better understanding of the sources and magnitudes of variability impacting terrestrial field dissipation studies could result in improved study design and conduct.

References

1. Pesticide Reregistration Rejection Rate Analysis: Environmental Fate; EPA 738-R-93-010; United States Environmental Protection Agency Washington, DC, 1993; p156.
2. Holzwarth, U., R.J. Blomer, and M. Uihlein. "Influence of Sampling Technique on Representative Sampling in Soil Dissipation Studies." Poster presented at the 8th IUPAC International Congress, Washington, DC, 1994.
3. Mojasevic, M., and C.S. Helling. 9th Int. Cong. Pestic. Chem., IUPAC, 1998, Abstr. No. 2:7C-049.
4. Woodrow, J.E., J.N. Seiber, and L.W. Baker. Environ. Sci. Technol. 1997, 31, 523.
5. Braverman, M.P., J.S. Corley, D.C. Thompson, M. Arsenovic, V.R. Starner, K.S. Samoil, F.P. Salzman, D.L. Kunkel, and J.J. Baron. Proc. Weed Sci. Soc. Amer., 2001, 41, 17.
6. TeeJet Spraying Systems Co. Agricultural Spray Products Catalog, 1998, 47, 134.
7. American Crop Protection Association. Workshop minutes,"Zero-time residue levels in field soil dissipation studies." 1996, Crystal City, VA.
8. Taylor, A.W., H.P. Freeman, and W.M. Edwards. J. Agric. Food Chem. 1971, 19, 832.
9. Wauchope, R.D., J.M. Chandler, and K.E. Savage. Weed Sci. 1977, 25, 193.
10. McWhorter, C.G. and O.B. Wooten. Weeds. 1961, 9, 42.
11. Nordby, A. and R. Skuterud. Weed Res. 1975, 14, 385.
12. Cross, R.F. and R.E. Majors. LC-GC Magazine. 2000, 18(5),468.
13. Trubey, R.K., R.A. Bethem and B. Peterson. J. Agric. Food Chem. 1998,46,2360.
14. Cline, M.G. Soil Sci. 1944, 58, 275.
15. Walker, A. and P. A. Brown. Crop Protection. 1983, 2(1), 17.

16. Vischetti, C., M. Businelli, M. Marini, E. Capri, M. Travisan, A.A.M. Del Re, L. Donnarumma, E. Conte, and G. Imbroglini. Pest. Sci. 1997,50,175.
17. Rao, P.S.C. and R.J. Wagenet. Weed Sci. 1985, 33(suppl. 2), 18.
18. Wood, L.S. H.D. Scott, D.B. Marx and T.L. Lavy. J. Environ. Qual. 1987, 16(3), 251.
19. Novak, J.M, T.B. Mooreman, and C.A. Cambardella. J. Environ. Qual. 1997, 26,1271.
20. Parkin, T.B. J.Environ. Qual. 1993, 22, 409.
21. Warrick, A.W. In *Environmental Soil Physics;* D. Hillel, ed.; Academic Press: New York, NY, 1998; 655.
22. Russo, D. and E. Bresler. Soil Sci. Soc. Am. J. 1981, 45, 682.
23. Parkin, T.B. Soil. Sci. Soc. Am. J. 1987, 51, 1194.
24. Parkin, T.B. and D.R. Shelton. J. Environ. Qual. 1992, 21, 672.
25. Vauclin, M., S.R. Viera, R. Bernard, and J.L. Hatfield. Water Res. Res. 1982, 18(6), 1677.
26. Nolin, M.C., S.P. Guertin, and C. Wang. In *Proceedings from the 3rd International Symposium on Precision Agriculture.* Minneapolis, MN. ASA/CSSA/SSSA. 1996, 257.
27. Rao, P.S.C., C.A. Bellin and M.L. Brusseau. In *Sorption and Degradation of Pesticides and Organic Chemicals in Soil.* D. Linn, ed.; Soil Sci. Soc. Amer.: Madison, WS, 1993; 260.
28. Lefrance, P. and Banton, O. Geoderma. 1995, 65, 331.

Chapter 7

Spatial Variability of Herbicide Sorption on Soil

W. C. Koskinen[1], D. J. Mulla[2], R. S. Oliveira, Jr.[3],
B. R. Khakural[2], and P. C. Robert[2]

[1]Agricultural Research Service, U.S. Department of Agriculture,
1991 Upper Buford Circle, Room 439, St. Paul, MN 55108
[2]Department of Soils, Water, and Climate, University of Minnesota,
1991 Upper Buford Circle, St. Paul, MN 55108
[3]Departamento de Agronomia, Universidade Estadual de Maringá,
Maringá, PR 86020–900, Brazil

A limitation in using sorption coefficients to predict herbicide transport is the spatial variability of soil properties over large fields. Spatial variability in alachlor and imazethapyr sorption was determined on samples from a 31.4-ha field: pH 4.9-7.6, 1.45-5.80% OC, and 26-65% clay. Alachlor sorption Kd ranged from 5.45 to 21.5. OC content was the most important property influencing sorption. Imazethapyr Kd varied from 0.18 to 3.78, but showed two distinct patterns in spatial distribution: areas with pH > 6.2 where Kd variation was based on pH; and areas with pH < 6.2, where Kd variation was also affected by OC. Based on spatial variability of soil properties and sorption, the field could be divided into management areas for site-specific herbicide application to reduce potential off-site transport. However, field-scale spatial patterns in sorption can vary with the method of interpolation. Linear sorption model based regression methods do not appear to adequately represent the spatial patterns of sorption in soil. It appears more sophisticated geostatistical approaches, such as co-kriging must be used.

Sorption is one of the most important processes impacting the fate of pesticides as it retards their movement through the soil profile. The magnitude and strength of sorption depend on the molecular characteristics of the pesticide and the soil chemical and physical properties. Sorption varies with a number of soil properties, but soil organic carbon (OC) content has been considered the single best predictor of sorption for nonpolar, nonionizable organic chemicals (1). For polar, ionizable chemicals such as weak bases like atrazine, factors such as pH, OC and clay contents affect sorption (2).

Soil physical and chemical properties in the field can vary both spatially and temporally. Intrinsic spatial variability includes natural variations in soil characteristics such as particle size and OC content, which might result from soil formation processes (3). The effect of spatial variability on pesticide fate has been discussed in the literature as a function of farming management practices (4) and as it affects sampling requirements (5), sorption (6), and transport (7,8).

Sorption is usually characterized by using the batch equilibration method to determine sorption distribution coefficients (Kd) or Freundlich coefficients (Kf). The value of Kd or Kf is a measure of the extent of the interaction of the pesticide with soil and is basic input for different research, screening, and management simulation models (9,10). The correlation between Kd and the organic carbon fraction of the soil for nonionizable herbicides allows estimation of specific sorption properties (Koc) for a herbicide largely independent of soil type (2). Most of the leachibility indices proposed (11,12) adopt Koc as the most important single value to rank the pesticides.

One of the significant limitations of a Kd value to predict solute transport through natural soils is spatial variability of soil properties over large field areas (13). Although Kd measurements have been successful in describing laboratory phenomena, these values have not provided an accurate tool to predict field behavior in some studies (14,15). This problem has been attributed to local variations in transport resulting from lateral and vertical variability not represented by a single average sorption coefficient (16,17). Prediction of pesticide transport by models may be improved by taking into account field variations in sorption. However, field sampling for model validation becomes a difficult task if a large number of samples must be taken to adequately account for spatial variability. In pesticide fate and efficiency studies, the problems associated with a large number of samples are particularly evident due to the high cost of pesticide residue analysis (3).

It is impractical to measure Kd values for the many soil-pesticide combinations of environmental interest because of the time and costs involved. Predicting pesticide sorption using easily measured soil properties has, therefore, been the focus of numerous research studies. A distinct potential advantage of this approach is that, given the variability of a more easily measured parameter such as soil OC, pH, or texture, the expected variability of a related parameter such as Kd, which is more difficult or expensive to measure, can be estimated. Ultimately, improving the knowledge of factors influencing sorption of herbicides in soil may provide useful guidelines to precision agriculture. For instance, site-specific application of alachlor spraying to a fine loamy soil decreased adverse environmental effects by reducing concentrations of the herbicide in surface runoff (18).

This paper evaluates imazethapyr and alachlor sorption within a 31 ha field, to (a) determine the most important soil properties influencing sorption across a landscape; (b) characterize spatial variability of sorption across this field; and (c) discuss criteria to separate field portions in such a way that imazethapyr and alachlor application could be modulated based on the correlation of sorption and soil properties.

Materials and Methods

Soil samples

Samples of surface soil (0-20 cm) were obtained from a 31.4-ha field farmed under a conventional corn-soybean rotation located at Blue Earth County, MN. The predominant soils were: Waldorf silty clay loam (SiCL) (fine, montomorillonitic, mesic Typic Haplaquoll), Lura SiCL (fine, montmorillonitic, mesic Cumulic Haplaquoll), Shorewood SiCL (fine, monmorillonitic, mesic Aquic Argiudoll), Blue Earth SiCL (fine-silty, mixed (calcareous) mesic, Mollic Furaquent), Lester loam (fine-loamy, mixed, mesic Mollic Hapludalf), and Cordova clay loam (fine-loamy, mixed, mesic Typic Argiaquoll). Nine south-north parallel transects 792.48 m long, each 45.72 m apart were established. Each transect was sampled at intervals of 30.48 m. Composite soil samples were taken from each grid point. The soil samples were air-dried and passed through a 2-mm sieve. Particle size distribution was performed using the hydrometer method, and soil organic matter content was determined using a modified Walkley Black method. The pH was determined in a 1:1 soil:water suspension.

From the total of 235 soil samples, 35 samples (Table I) were used to determine the spatial field variability of the sorption coefficients (K_d) values in the field and to determine the relationship between K_d and soil characteristics. The 35 soil samples were selected to be representative of the entire field; the mean, range, and standard deviations of the values of physical and chemical properties of the 35 soil samples were similar to those of the 235 samples.

Table I. Descriptive statistics for soil characteristics for the whole field and selected soil samples

	Field (n=235)			
	Mean	*Range*	*SD[a]*	*CV*
pH	6.24	4.9-7.7	0.8	12.9
OC (g kg^{-1})	4.09	1.28-5.80	1.20	29.4
Clay (%)	42.0	25.6-65.1	5.50	13.1
Silt (%)	43.8	21.6-58.9	7.20	16.5
Sand (%)	14.3	1.10-48.3	8.13	57.0
	Selected samples (n=35)			
pH	6.19	4.9-7.6	0.8	12.8
OC (g kg^{-1})	3.71	1.45-5.80	1.26	33.9
Clay (%)	42.0	25.6-58.9	5.81	13.9
Silt (%)	43.6	26.2-58.9	6.94	15.9
Sand (%)	14.6	2.09-48.3	9.58	65.8

[a] Abbreviations: SD, standard deviation; CV, coefficient of variation.

Chemicals

Imazethapyr (99% purity) was purchased from Chem-Service and uniformly labeled ^{14}C-imazethapyr (97.5% radiopurity, specific activity 214 kBq nmol^{-1}) was obtained from American Cyanamid. Alachlor (98% purity) was purchased from Chem-Service and uniformly labeled ^{14}C-alachlor (>99% radiopurity; specific activity 999 kBq nmol^{-1}; initial radioactivity in solution of 86.0 KBq L^{-1}) was obtained from Pathfinder Laboratories.

Sorption-desorption studies

Batch equilibration sorption measurements were conducted with quadruplicate samples, by weighing 5-g subsamples into 50-mL glass centrifuge tubes and adding 10 mL of a solution of 0.3456 µmol L^{-1} imazethapyr in 0.01 *M* CaCl$_2$, with ~ 70 Bq mL^{-1} ^{14}C-imazethapyr or by weighing 10-g subsamples into 50-mL glass centrifuge tubes and adding 10 mL of solution of ^{14}C-alachlor (5.56 µmol L^{-1} 0.01 *M* CaCl$_2$; ~ 70 Bq mL^{-1} ^{14}C). The tubes were then capped with teflon-lined screw caps and shaken for 24 h. The soil slurries were centrifuged for 30 min at 2000 rpm. Five mL of the supernatant were removed and placed into a vial. Five mL of scintillation cocktail were added to a 1-mL aliquot of the supernatant and the ^{14}C quantified by liquid scintillation counting (LSC) using a Packard 1500 Tri-carb liquid scintillation analyzer.

Subsequently, four soil samples with similar OC contents and different pH values were used for a sorption-desorption experiment with initial ^{14}C-imazethapyr solution concentrations of 0.035, 0.173, 0.691, and 3.46 μmol L^{-1}. Four soils with different OC contents were equilibrated with initial ^{14}C-alachlor solution concentrations of 0.45, 1.34, 4.45, and 11.14 μmol L^{-1}. All solutions were prepared in 0.01M CaCl$_2$, with ~ 70 Bq mL^{-1} [14]C, and batch sorption equilibration measurements were performed as previously described.

Desorption was determined by sequentially replacing the supernatant removed with the same amount of 0.01 M CaCl$_2$, resuspending the soil slurry, and reequilibrating it for 24 h. Desorption isotherms were obtained using three successive desorption cycles. Imazethapyr and alachlor concentrations in the desorption solutions was determined and the amount of herbicide that remained sorbed on the soil after each desorption step was calculated by difference.

Data analysis

Statistical analyses were performed using the statistical package SigmaStat-SigmaPlot. Means, standard deviations, minimum and maximum values, and coefficients of variation were determined on all properties. Pearson product moment correlation and multiple linear regression analyses were performed to determine the correlation between K_d and soil properties. A map showing the spatial distribution of K_d values was prepared using a kriging procedure (19).

Sorption data were fit to the logarithmic form of the Freundlich equation, log Cs = log K_f + 1/n log Ce, where Cs (μmol kg^{-1}) is the amount of herbicide sorbed by soil, Ce (μmol L^{-1}) is the concentration in solution, and K_f and 1/n are empirical constants representing the intercept and the slope of the isotherm, respectively. The Freundlich coefficients normalized for organic carbon content ($K_{f,oc}$) were calculated by using the equation $K_{f,oc} = (K_f/OC\%) \times 100$.

Results and Discussion

Imazethapyr

Imazethapyr sorption data fit the Freundlich equation ($r^2 \geq 0.998$) In general, sorption decreased with increasing soil solution pH, as indicated by K_f

values (Table II). Because sorption was not concentration dependent, $1/n$ values ranged from 0.98 to 1.00, K_d values were used to determine spatial variability in soil at the selected field site. Individual measurements of K_d varied from 0.18 to 3.78 across the field with an average value of 1.56. The dominant soil characteristic correlated to K_d for the whole field was pH ($r^2 = 0.83$).

Table II. Freundlich sorption parameters for imazethapyr as a function of soil pH on samples with 4.29 g kg^{-1} OC

pH	K_f	$K_{f,oc}$	$1/n$ sorption	r^2	$1/n$ desorption	r^2
	$\mu mol^{(1-1/n)}L^{1/n}kg^{-1}$					
5.5	2.45	57	0.98	0.999	0.19	0.847
	(2.28-2.63)[a]	(53-61)	(±0.02)		(±0.06)	
5.7	2.08	48	1.00	0.999	0.19	0.870
	(1.95-2.21)	(45-52)	(±0.02)		(±0.05)	
6.2	0.61	14	0.99	0.998	0.20	0.849
	(0.57-0.65)	(13-15)	(±0.03)		±0.06)	
7.3	0.30	7	1.00	0.999	0.22	0.890
	(0.29-0.31)	(7-7)	(±0.01)		(±0.05)	

[a]Numbers in parenthesis are confidence intervals (K_f, $K_{f,oc}$) or standard deviation of the mean ($1/n$), (adapted from 20)

Imazethapyr is an amphoteric molecule, having both carboxylic acid and pyridine functional groups. Dissociation to the anion results in very low sorption at high pH levels. The nonionized form of acidic herbicides, predominant at pH below its pKa of 3.9, exhibits binding more typical of nonionic herbicides. At low pH, imazethapyr can be protonated and sorbed on the soil by cation exchange (21,22). As a result, the soil behavior of imidazolinone herbicides, such as imazethapyr, has been shown to be affected by pH (23), OC content (24), and mineralogy (25). In our study, there were also significant correlations (P < 0.05) between K_d and silt content and between K_d and sand content, however these properties had no impact on the prediction of K_d from soil properties. There was no significant correlation (P > 0.05) between K_d and clay or K_d and OC content for the whole field. Significant negative correlations between pH and K_d have also been described previously (23).

The analysis of K_d values showed two distinct patterns in spatial distribution (below and above pH 6.2), and this was used as a criteria to divide this field into two potential management areas (Figure 1): areas where pH > 6.2 and K_d< 1.5, and areas of pH < 6.2 and K_d > 1.5. In the high pH area, K_d variation was based

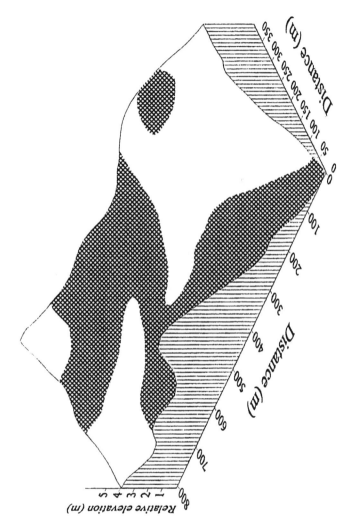

Figure 1. Spatial distribution of Kd showing distinct patterns for areas with higher (cross-hatch, Kd < 1.5) or lower (white, Kd > 1.5 leaching potential of imazethapyr.

primarily on pH variation; whereas in the low pH area, soil OC also had a significant influence on K_d variation. This separation allowed us to identify portions of the field where herbicide sorption would be minimal, with high potential for leaching losses (i.e. areas with $K_d < 1.5$).

Predictive equations based on pH and OC contents can be calculated, i.e $K_d = 9.37 - 1.29$ pH ($r^2 = 0.83$) (for the whole field), $K_d = 3.94 - 0.488$ pH ($r^2 = 0.78$) (for areas with $K_d < 1.5$), or $K_d = 7.68 - 0.867$ pH $- 0.103$ OC ($r^2 = 0.55$) (for areas with $K_d > 1.5$). However, more precise equations to estimate K_d value for imazethapyr are needed as a function of specific soil characteristics that affect its behavior. For instance, these predictive equations are only based on sorption coefficients and do not include desorption characteristics. Desorption isotherm slopes ($1/n$ - desorption) were independent of concentration and their values were much lower than those for the sorption isotherms, indicating hysteresis occurred during desorption of imazethapyr (Table II). Therefore once sorbed, imazethapyr does not readily desorb.

Use of these data in simulation modeling could provide a general parameter about the intensity and probability of imazethapyr leaching in soil profile in different areas of this field. Lafrance and Banton (*26*) have previously performed a set of simulations using the VULPEST model to evaluate the impact of spatial variability of OC in controlling the leaching of herbicides through soil. Through simulation techniques, one would be able to estimate the maximum rate of imazethapyr that should be applied to each of the proposed management zones to avoid excessive leaching in the more vulnerable areas. Ultimately, this could lead to the recommendation of site-specific rates for this field plot.

Alachlor

Measured *Kd* values on the representative samples ranged from 5.45 to 21.5 L kg^{-1}. Pearson correlation coefficients indicated that OC was the single most important soil variable affecting alachlor sorption ($r = 0.85$), although % clay was also significantly correlated to sorption. After normalizing the observed *Kd* values for the OC content, no significant relationships ($P > 0.05$) were obtained when *Koc* was correlated to soil properties. *Koc* for the 35 samples had a mean of 285 and a range from 173 to 563. These data confirm the well-documented relationship between alachlor sorption and soil organic carbon (i.e., *27,28*) and, to a lesser extent, mineral soil components (*29*). Clay:OC ratios in samples varied from 7 to 25, with smaller *Kd* values at the higher ratios. It has been suggested that H-binding and charge-transfer processes are the major mechanisms occurring at low concentrations, and a hydrophobic binding of alachlor to the aliphatic parts of humic is dominant at high concentrations (*30*).

Data for alachlor sorption on four soils with similar textures, but increasing OC, fit the Freundlich equation ($r^2 \geq 0.99$) (Table III). The slope ($1/n$) of sorption isotherm was < 1.0 on soils with OC contents < 4.8% indicating that as the concentration of the herbicide in solution increased, the percent of soil-sorbed herbicide decreased. Although the Freundlich coefficients (Kf or Kf,oc and $1/n$) would more accurately characterize alachlor sorption, for simplicity and because $1/n > 0.90$, the following models were developed with OC and clay content for predictions of alachlor sorption: $Kd = 0.454 + 2.70$ OC ($r^2 = 0.725$) and $Kd = -1.25 + 2.58$ OC $+ 0.0513$ clay ($r^2 = 0.729$). The combination of OC and clay (%) in multiple regression analysis did not significantly improve the prediction of Kd compared to using OC alone.

Table III. Freundlich sorption and desorption parameters of alachlor in soil samples with different organic carbon contents

OC Content	Kf^{a}	Kf,oc^{b}	Sorption		Desorption	
			$1/n$	r^{2c}	$1/n$	r^{2c}
%	$\mu mol^{(1-1/n)} L^{1/n} kg^{-1}$					
2.20	5.29	240	0.92	0.99	0.28	0.97
	(5.24-5.34)	(238-243)	(±0.01)		(±0.04)	
3.89	8.74	225	0.91	0.99	0.13	0.95
	(8.59-8.88)	(221-229)	(±0.01)		(±0.02)	
4.81	11.26	234	0.91	0.99	0.11	0.86
	(11.18-11.34)	(232-235)	(±0.01)		(±0.03)	
5.80	12.89	222	0.97	0.99	0.13	0.95
	(12.70-13.09)	(219-226)	(±0.01)		±0.02)	

[a]Numbers in parenthesis are standard deviation of the mean.

[b]Kf corrected for organic carbon content.

[c]Correlation coefficients of linearized Freundlich isotherms.

Alachlor desorption exhibited pronounced hysteresis (desorption $1/n$ < sorption $1/n$ in all four soil samples and at all initial sorbed amounts of alachlor. The degree of irreversibility in the sorption-desorption process increased with an increase in sorbed alachlor. Hysteretic desorption has been described previously (29,31). Greater hystersis occurred in soils with the highest OC contents (Table III), demonstrating that the organic matter also impacts the desorption process. In spite of these limitations, the linear relation between Kd and soil OC was chosen to characterize the spatial variability of alachlor sorption.

Geostatistical approaches have also been utilized in defining site-specific fertilizer recommendations (*32*), varying application within a field on the basis of variability of soil properties. This technique may also provide some help in defining site-specific pesticide rates. Using measured values of OC and the calculated relationship between alachlor sorption *Kd* and OC content, a linear sorption model based regression method could be used to represent the spatial patterns of sorption in soil (Figure 2). The separation allows an overview of high/low OC contents and, therefore, of alachlor sorption to soil. One of the criteria proposed by Environmental Protection Agency (EPA) (*33*) to trigger restrictions on pesticides that may leach to groundwater is a *Kd* of less than 5 L kg^{-1}. Based on the above equation, this *Kd* value corresponds to an OC content of 1.69%. This value can be used as an indication to separate field portions with the greatest potential to allow excessive leaching of alachlor, which is a relatively small area of the field. Whether this would be the best approach is unclear. A cokriging procedure, whereby sorption *Kd* values were obtained by cokriging interpolation of measured *Kd* values using measured OC as a covariate, resulted in completely different spatial representation of *Kd* (Figure 3). It appears that more sophisticated geostatistical approaches, such as cokriging, must be used to characterize spatial variation in soil properties and processes that impact herbicide behavior in soil.

Summary

Availability of soil-applied pesticides to control weeds and for transport is directly controlled by the strength and ratio of binding to soil. Reducing the rates of application through site-specific rates is a potential mitigation method for reducing herbicide concentrations in surface and groundwater. In some cases, where decreasing rates may impair satisfactory weed control, band application or alternative weed control methods may be required. But to stipulate a specific rate to be applied to certain portions of the field is not a simple task, and demands extensive field validation, since the efficacy of a herbicide depends on non-controllable variables, such as climatic conditions, weed species, and their susceptibility to the specific herbicide. Clay et al. (*28*) suggested that the control of sorption and degradation processes in combination with good management may be sufficient to prevent alachlor contamination of an aquifer. Forcella (*34*) suggested that management of spatial variability is worthwhile as long as the degree of variability is large enough to justify the cost of obtaining information and managing the differences accordingly.

We have demonstrated that imazethapyr sorption to soil is primarily affected by soil pH, but can also be influenced by other soil characteristics, such as OC

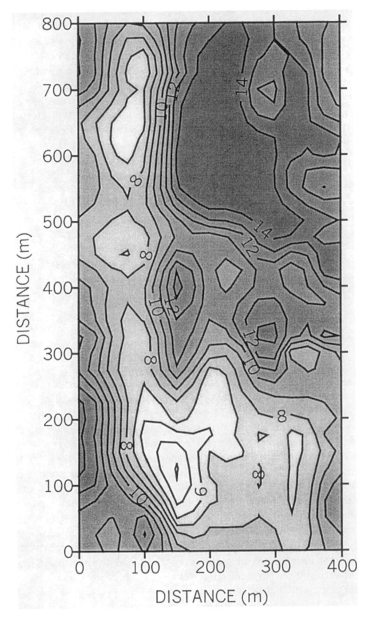

Figure 2. Alachlor Kd values from linear sorption model based on OC and Koc for the watershed.

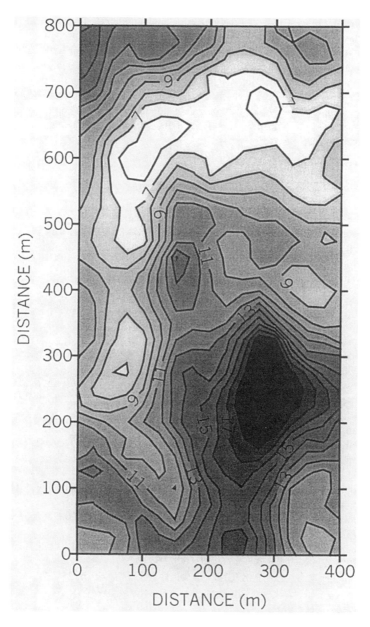

Figure 3. Kriged alachlor Kd values for the watershed.

contents under acidic conditions, while alachlor sorption is primarily affected by OC content. Our research also demonstrates how simple equations based on soil properties can be used to estimate sorption coefficients, and coupled with simulation techniques, to predict site-specific herbicide behavior within a field. While the division of a field based on soil properties correlated to sorption provides a rationale for site-specific management, specific application rates however, will depend on a more complex analysis involving the level of weed infestation and distribution, availability of adequate spraying equipment, and economic analysis. The division of the field will also be dependent on the statistical approach used.

Acknowledgements

The authors wish to express their appreciation to Coordenadoria de Aperfeiçoamento do Pessoal de Nível Superior, CAPES (Brazil) by providing financial support to this research and American Cyanamid Co. for the ^{14}C-labeled imazethapyr.

Literature Cited

1. Koskinen, W. C.; Harper, S. S. In *Pesticides in the Soil Environment: Processes, Impacts, and Modeling;* Cheng,, H. H., Ed. SSSA: Madison, WI, 1990; pp 51-77.
2. Green, R. E.; Karickhoff, S. W. In *Pesticides in the Soil Environment: Processes, Impacts, and Modeling;* Cheng,, H. H., Ed. SSSA: Madison, WI, 990; pp 79-101.
3. Rao, P. S. C.; Wagenet, R. J. *Weed Sci.* **1985**, *33(suppl. 2)*, 18-24.
4. Mallawatantri, A. P.; Mulla, D. J. *J. Environ. Qual.* **1992**, *21*, 546-551.
5. Smith, C. N.; Parrish, R. S.; Carsel, R. F. *Environ. Toxicol. Chem.* **1987**, *6*, 343-357.
6. Novak, J. M.; Moorman, T.B.; Cambardella, C. A. *J. Environ. Qual.* **1997**, *26*, 1271-1277.
7. Amoozegard-Fard, A.; Nielsen, D. R.;Warrick, A. W. *Soil Sci. Soc. Am. J.*, **1982**, *46*, 3-9.
8. Chammas, G. A,; Hutson, J. L.; Hart, J. L.; DiTomaso, J. M. *Weed Technol.* **1997**, *11*, 98-104.
9. Wagenet, R. J.; Rao, P. S. C. *Weed Sci.* **1985**, *33(Suppl. 2)*, 25-32.
10. Brown, C.D.;Baer, U.; Günther, P.; Trevisan, M.; Walker, A. *Pestic. Sci.* **1996**, *47*, 249-258.

11. Jury, W. A.; Focht, D. D.; Farmer, W. F. *J. Environ. Qual.* **1987**, *16*, 422-428.

12. Gustafson, D. I. *Environ. Toxicol. Chem.* 1989, **8, 339-357.**

13. Elabd, H.; Jury, W. A.; Cliath, M. M. *Environ. Sci. Technol.* **1986**, *20*, 256-260.

14. Rao, P. S. C.; Green, R. E.; Balasubramanian, V.; Kanehiro, Y. *J. Environ. Qual.* **1974**, *3*, 197-202.

15. Flury, M. *J. Environ. Qual.* **1996**, *25*, 25-45.

16. Van de Pol, R. M.; Wierenga, P. J.; D. R. Nielsen, R. R. *Soil Sci. Soc. Am. J.* **1977**, *41*, 10-13.

17. Jury, W. A.; Stolzy, L. H.; Shouse, P. *Water Resour. Res.* **1982**, *18*, 369-375.

18. Khakural, B. R., Robert, P. C., and Koskinen, W. C. 1994 Soil Use Manage. 10:158-164.

19. Golden Software. Surfer for Windows, Version 6. Golden Software Co.: Golden , CO, 1996.

20. Oliveira Jr, R. S.; Koskinen, W. C.; Ferreira, F. A.; Khakural, B. R.; Mulla, D. J.; Robert, P. C. *Weed Sci.* **1999**, *47*, 243-248.

21. Stougaard, R. N.; Shea, P. J.; Martin, A. R. *Weed Sci.* **1990**, *38*, 67-73.

22. Bresnahan, G. A.; Koskinen, W. C.; Dexter, A. G.; Lueschen, W. E. *J Agric. Food Chem.* **2000**, *48*, 1929-1934

23. Renner, K. A.; Meggit, W. F.; Penner, D. *Weed Sci.* **1988**, *36*, 8-83.

24. Goetz, A. J.; Lavy, T. L.; Gbur Jr, E. E. *Weed Sci.* **1990**, *38*, 421-428.

25. Che, M.; Loux, M. M.; Traina, S. J.; Logan, T. J. *J. Environ. Qual.* **1992**, *21*, 698-703.

26. Lafrance, P.; Banton, O. 1995. Pages 125-132 *in* K. Kovak and J. Krásný, eds. Groundwater Quality: Remediation and Protection (Proceedings of the Prague Conference). IAHS Publ. No. 225. Wallingford, Oxfordshire, UK: IAHS Press, Institute of Hydrology.

27. Peter, C. J.; Weber, J. B. *Weed Sci.* **1985,** 874-881.

28. Clay, S. A.; Moorman, T. B.; Clay, D. E.; Scholes, K. A. *J. Environ. Qual.* **1997**, *26*, 1348-1353.

29. Locke, M. A. *J. Environ. Qual.* **1992**, *21*, 558-566.

30. Senesi, N.; Brunetti, G.; La Cava, P.; Miano, T. M. *Soil Sci,* **1994**, 176-184.

31. Yen, P. Y.; Koskinen W. C.; Schweizer E. E. *Weed Sci.* **1994,** *42*, 233-240.

32. Mulla, D. J. In *Soil Specific Crop Management - A Workshop in Research and Development Issues,* Roberts, P. C., Ed, ASA, CSSA, and SSSA: Madison, WI, 1993; pp 15-23.

33. Cohen, S. Agricultural chemical news. *Ground Water Monitor. Remed. 1991,* 79-80.

34. Forcella, F. In *Soil Specific Crop Management.* P. C. Robert, P. C.,Ed. ASA, CSSA and SSSA: Madison, WI, 1993; pp 125-132.

Chapter 8

Effect of Variability of Soil Properties as a Function of Depth on Pesticide Sorption–Desorption

Sharon A. Clay[1] and William C. Koskinen[2]

[1]Department of Plant Sciences, South Dakota State University,
Brookings, SD 57007
[2]Agricultural Research Service, U.S. Department of Agriculture,
1991 Upper Buford Circle, Room 439, St. Paul, MN 55108

Sorption-desorption is arguably the most important process affecting the transport of pesticides through soil since it controls the amount of chemical available for transport. Sorption is usually characterized by determining surface soil sorption coefficients that are then used in solute transport models. Significant drawbacks to using surface soil sorption coefficients to predict pesticide transport are the spatial variability of 1) surface soil properties over large fields and 2) soil properties in the soil profile. Our objective is to give an overview of how pesticide sorption and desorption is influenced by changes in soil properties with depth. This will be illustrated using several classes of pesticides. Specific pesticides include atrazine, alachlor, sulfometuron methyl, tebuthiuron, and imidacloprid. Results indicate that correlations between pesticide sorption-desorption and soil properties of surface soils cannot necessarily be used to characterize pesticide sorption-desorption in subsurface soils.

Pesticide sorption or retention to soil controls the amount available for pest control (1), microbial transformations, and transport (2). Spatial variability of sorption in the soil surface can be used to optimize pesticide rates throughout a field for pest control (1) and minimize rates where runoff (3) and carryover problems may occur. Sorption coefficients are used as basic inputs into pesticide transport research screening and management simulation models (4) such as LEACHP, PRZM-2, and VARLEACH (5) to quantify pesticide transport. Spatial variation of pesticide sorption due to soil heterogeneity as a function of depth

could be included in these models to better assess the probability of groundwater contamination at subfield, field, watershed, and /or regional levels (6, 7, 8).
The intensity and extent of soil sorption depends on the pesticide molecular characteristics and soil chemical and physical properties with the amount sorbed ranging from 0 to 100%.

Pesticide sorption often is characterized by batch equilibration methods to determine the sorption partition coefficient (Kd) (using a single pesticide concentration) or Freundlich coefficients (Kf) (using several pesticide concentrations). The Kd and Kf values are ratios of the amount of pesticide sorbed to soil compared to that in solution after equilibration.

Desorption is the measure of the amount of sorbed chemical that can be removed from soil and can range from 0 to 100%, depending on soil and chemical characteristics. Desorption often is measured in the same batch equilibrium studies immediately after sorption equilibration has been completed. Unfortunately, in a majority of cases desorption cannot be predicted from the sorption equilibrium and hysteresis occurs (9, 10). Desorption is complicated further when chemical residues are allowed to age in soil. Normally as the aging process continues, even less chemical is removed (11, 12). Some of the chemical can be recovered after exhaustive extraction, while some is irreversibly bound to soil components (13). When modeling pesticide residue movement in soil, the amount that can be removed easily is most likely what will contribute to pesticide loading in groundwater.

Effects of Pesticide Characteristics on Sorption

The chemical properties of the pesticide can greatly influence pesticide sorption-desorption. These properties include the ionic nature of the molecule, solubility, polarity, and the type and number of functional groups on the molecule. For instance, cationic pesticides, like glyphosate and paraquat, are bound to soil by cation ion exchange on clay minerals (14) and are difficult to desorb from clays by exchange with inorganic cations.

Ionizable compounds such as basic compounds (triazines and pyridinones) and acidic compounds (carboxylic acids and phenols) can sorb by ionic mechanisms when they are ionized. Weakly basic compounds may sorb by cation exchange; weakly acidic compounds may sorb by anion exchange. For these chemicals, ion exchange is not the sole sorption mechanism; other physicochemical forces can also be involved.

Depending on the pH of the soil-water system, weakly acidic pesticides exist as either the undissociated molecule or the corresponding anion. Although anionic pesticides can be sorbed by anion exchange, it is not likely as clays and soil organic carbon (OC) (made up of plant residue, humus, and living organisms) are either negatively charged which would repel anionic species or noncharged. Therefore, most soils sorb no or minimal amounts of anionic pesticides. The amount of anionic pesticide desorbed (if any amount is sorbed) is usually quite high (15).

Weakly basic pesticides, such as triazines (i.e atrazine) and pyridinones (i.e fluradone), can be easily protonated at low soil pH levels. In most soils, these chemicals exist as both cationic and molecular species, therefore a variety of sorption mechanisms are operative. Electrostatic interactions can also be important because attraction can develop between the polar molecule and heterogeneous soil surfaces that have ionic and polar sites, resulting in stronger bonds. Kd is affected by OC and clay content as well as soil pH. Desorption of weak bases from soil has been shown to be hysteretic with amounts from 10 to 90% of the amount applied and generally decreasing if chemicals are allowed to age in soil.

Sorption of nonpolar, nonionizable pesticides occurs by weak attractive interactions such as van der Waals forces. Sorption of these pesticides also are affected by clay and OC content. Generally, the greater the amount of clay and OC content, the more pesticide is retained. Desorption of this class of chemicals often is dependent on the amount sorbed, soil texture, and OC content. Hysteresis is a common phenomenon and varies with soil horizon.

Effect of Soil Properties on Sorption-Desorption

Pesticides are sorbed on both inorganic and organic soil constituents by a variety of mechanisms. The major soil components that sorb pesticides are soil clay and OC (15). Clay particles can be made up of several clay minerals (vermiculite, smectite, kaolinite and others), each having a distinct ability to sorb pesticides (14, 16). Other soil properties that can influence pesticide retention include soil pH, oxide content (iron and aluminum) (17, 18, 19), and sand content (usually negatively correlated with pesticide sorption). The relative importance of organic and inorganic constituents on pesticide sorption depends on the chemical properties of the pesticide, and the amount, distribution, and properties of these constituents.

Clay and Soil OC content as function of depth. Clay and soil OC concentration and depth of maximum concentration vary considerably with soil type or series. However, general trends in clay and OC content with depth can be described if the soil order (the broadest soil taxonomic category) is known (20).

Alfisols and Mollisols are two of the most important soil orders to agriculture in the midwestern United States. Alfisols normally develop under deciduous forests. Silicate clay accumulates by illuviation with clay skins or films present in the B horizon. The cation exchange capacity (CEC) of Alfisols is more than 35% saturated with base-forming cations such as calcium and magnesium. The OC content of the epipedon (diagnostic horizons that occur at the soil surface) may be greater than 0.6%, but have a base saturation of <50% (umbric epipedon) or have an OC content of <0.6% (ochric epipedon).

Mollisols develop under grassland and are characterized by the accumulation and abundance of OC in the surface soil. This soil order is characterized by a mollic epipedon that has OC content >0.6% that is generally >25 cm in thickness with a high (>50%) base saturation.

In most agricultural soils, OC is greatest in the surface horizon due to the addition of plant residues over time. The correlation of sorption to OC content is often considered to be the single best predictor of sorption. The Koc [(Kd*100)/%OC] (21) has been substituted for Kd in some predictive transport models (22, 23). However, using Koc to predict pesticide leaching may be erroneous for several reasons. Sorption may be correlated to Koc but the correlation coefficient may range from 0.1 to 0.9 depending on soil type. This may lead to Koc values that are more variable than the original Kd values (7). Koc values also are not stable but change with depth and quality of OC present. Seybold et al. (24) reported differences in surface Koc values of similar soil types with similar OC content. Higher Koc values were calculated for soil developed under forest vegetation than soil developed under prairie vegetation. This led to lower atrazine mobility in the soils originally in forest. Another consideration is that many other factors may control sorption, especially in oxidized soils and at lower depths.

Soil pH as function of depth. The pH of a soil can vary greatly with depth but generally increases with depth in the profile because the surface is more weathered. If carbonates are present, and the soil is young, unweathered, and well drained, the soil pH may be relatively uniform throughout the profile.

Human activities can bring about large changes in surface soil pH (25, 26). The amount and type of fertilizer applied can drastically change soil pH over a short (days or weeks) or long duration (years). For example, ammonia applications will increase the surface pH to over 9 for weeks after application with a subsequent pH decrease. Ammonium sulfate or sulfur applications over a period of years can reduce the surface soil pH to about 4. Liming low pH soils will increase pH with a goal of raising the pH to about 7 in the top 15 cm. These changes in pH have been shown to influence sorption of weakly basic pesticides (27, 28, 29, 30).

Change in Sorption Characteristics as a Function of Depth

Pesticide adsorption. A significant limitation in using sorption coefficients to predict pesticide transport is the spatial variability and heterogeneity of soil properties in the soil profile. Representative examples of different classes of pesticides, i.e. atrazine, alachlor, sulfometuron methyl, tebuthiuron, and imidacloprid are presented below.

Benchmark agricultural soils of South Dakota were sampled and separated by horizon to determine atrazine and alachlor sorption isotherms (31). While

Table I. The measured vs predicted atrazine K_f values if K_{oc} of the A horizon is assumed to be constant for other horizons. Predicted K_f values = [K_{oc} (A horizon) * OC for the horizon of interest]/100.

Soil type*	Horizon	Depth	OC	K_f Measured	Predicted	K_{oc}
		(cm)				
Clarno	Ap	0-23	1.51	2.80		185.4
loam	Bw	23-41	1.10	1.57	2.04	142.7
	2C	91-152	0.05	0.61	0.09	1220
Moody	Ap	0-18	2.50	3.72		148.8
silty clay	Bw	18-43	0.55	0.92	0.82	167.3
loam	Bk	76-107	0.27	0.63	0.40	233.3
	2C	107-142	0.22	0.56	0.33	254.5
Nora	Ap	0-18	2.02	2.16		106.9
silty clay	Bw	18-46	0.51	1.02	0.55	200.0
loam	Bk	46-76	0.24	0.65	0.26	270.8
Brandt	Ap	0-41	4.50	5.35		118.9
silty clay	Bw	41-130	1.00	1.06	1.19	106.0
loam	2C	130-176	0.20	0.12	0.24	60.0

Adapted with permission from reference 31. South Dakota Acad. Sci., copyright 1997.
*Clarno loam is a fine-loamy, mixed, superactive mesic Typic Haplustoll; Moody and Nora silty clay loams are fine-silty, mixed superactive mesic Udic Haplostolls; Brandt silty clay loam is a fine-silty, mixed superactive frigid Calcic Hapludoll.

variability of atrazine and alachlor sorption within a soil type was not specifically characterized, the problem with assuming a constant Koc throughout the entire profile can be illustrated using these data (Tables I and II). In each of the soils, OC content decreased as depth in the soil profile increased. The Koc values ranged from 100 to 1200. However, if Koc values for the A horizon were assumed to represent the profile, then predicted vs measured Kf values ranged from 93% greater than the atrazine Kf in the Clarno Bw horizon (Table I) to 65% less than the alachlor Kf in the Nora Bk horizon (Table II). Overestimation of sorption would result in an underestimation of leaching potential, whereas an underestimation of sorption would result in overestimation of leaching potential.

Similar results have been reported for atrazine in sandy loam and loamy sand soils of Denmark (32). Koc values were similar through the top 80 cm but underestimated Kd from 50 to 100% at deeper soil profile depths.

Another example of inconsistent Koc values can be illustrated with the insecticide imidacloprid (a nitroguanadine) (Table III) and the degradation product guanadine (authors' unpublished data). These data came from two soil cores from the same soil series at the same landscape position. In the core designated as PZ, imidacloprid Koc values ranged from 181 at the 0 to 30 cm depth to 512 in the 671- to 793-cm depth, a 2.8 fold increase whereas the Kf value decreased by about 85%. The Koc values in the EB core are even more dramatic. The Koc of

Table II. The measured vs predicted alachlor K_f values if K_{oc} of the A horizon is assumed to be constant for other horizons. Predicted K_f values = $[K_{oc}$ (A horizon) * OC for the horizon of interest]/100.

Soil type	Horizon	Depth	OC	K_f Measured	K_f Predicted	K_{oc}
		(cm)				
Clarno	Ap	0-23	1.51	1.95		129.1
Loam	Bw	23-41	1.10	1.64	1.42	149.1
	2C	91-152	0.05	0.77	0.06	1540
Moody	Ap	0-18	2.50	3.36		134.4
silty clay	Bw	18-43	0.55	1.63	0.74	296.4
Loam	Bk	76-107	0.27	1.14	0.36	422.2
Nora	Ap	0-18	2.02	2.72		134.7
silty clay	Bw	18-46	0.51	1.55	0.69	303.9
Loam	Bk	46-76	0.24	1.18	0.32	491.7
Brandt	Ap	0-41	4.50	5.78		128.4
silty clay	Bw	41-130	1.00	1.88	1.28	188.0
Loam	2C	130-176	0.20	0.43	0.26	215.0

Adapted with permission from reference 31. South Dakota Acad. Sci., copyright 1997.

Table III. The measured vs predicted K_f values of imidacloprid if K_{oc} of the A horizon is assumed to be constant for other horizons. Predicted K_f values = [K_{oc} (A horizon) * OC for the horizon of interest]/100.

Core	Soil*	Depth (cm)	Clay (%)	OC (%)	K_{oc}	K_f Measured	Predicted	$1/n_{ads}$	$1/n_{des}$
PZ	SL	0-30	12.8	0.59	181	1.07		0.96	0.24
	SL	30-74	10.8	0.44	222	0.96	0.8	0.96	0.64
	SL	74-97	10.8	0.56	211	1.19	1.01	0.98	0.22
	SL	97-132	16.8	0.87	417	3.62	1.57	0.95	0.13
	LS	305-247	4.8	0.06	361	0.23	0.11	1.04	0.67
	S	671-793	2.8	0.03	512	0.15	0.05	1.03	0.7
EB	SL	0-30	10.8	0.41	282	1.15		0.94	0.29
	SL	30-74	10.8	0.22	457	1.01	0.62	0.9	0.33
	SiCL	74-97	38.8	0.85	289	2.45	2.39	0.95	0.27
	SL	97-132	20.8	0.44	405	1.78	1.24	0.91	0.28
	LS	305-427	4.8	0.09	279	0.26	0.25	0.96	0.46
	LS	671-793	4.8	0.01	10465	0.61	0.28	0.92	0.38

*SL = sandy loam, LS = loamy sand, S = sand, SiCL = silty clay loam, SiL = silt loam

imidaclorprid was 282 in the surface and increased to 10500 at the lowest depth, a 37-fold increase whereas the Kf value decreased by about 50%. If the Koc value of the surface increment is used, Kf values were underestimated from 10 to 70%. These data indicate that the mobility of the compound would be overestimated. Guanidine Koc values exhibited the same trend (data not shown).

The Kd and Koc sorption coefficients for sulfometuron methyl, tebuthiuron, and hexazinone were measured and calculated, respectively, for 6 sandy soils from upper Minnesota, Wisconsin, and Michigan (33). Results from three soils are presented in Table IV. Depending on the chemical and soil depth, Kd predicted in lower horizons from surface soil Koc values generally underestimated Kd especially at the 90- to 95-cm depth.

The OC content normally is very low at deep depths in the soil profile resulting in large Koc values if any sorption occurs. In addition, sorption to OC may not be the prominent component controlling sorption. Instead, sorption may be controlled by the presence of oxides (16, 17, 18) or other prominent physical or chemical processes (16).

Another consideration is that most sorption isotherms do not have a slope of 1. This indicates that a greater percentage of pesticide is sorbed at lower concentrations. For example, Felding (32) reported that Kd values were two times greater when the starting batch equilibrium concentration was 1.6 μg L^{-1} compared to Kd values calculated when the starting concentration was 5000 μg L^{-1}. Since pesticide concentrations are much lower below the plow layer, this effect may influence predictions of pesticide transport (2).

Another problem that can be present in the vertical profile is discontinuities of soil texture (e.g. sand lens, increases in OC, etc.) that may be present at inconsistent depths throughout a field. The imidacloprid example (Table III) illustrates this concept. The PZ core had a high OC content at 97- to 132-cm depth whereas the EB core had a higher OC content at 74- to 97-cm depth. Sorption in these areas increased, although Koc values were not necessarily correlated to the Kf values.

Pesticide desorption. The desorption process is very important in the prediction of pesticide movement in soils. This measurement provides an idea of the amount of pesticide that could be released from the depth of interest over a given time period. The amount of chemical desorbed is more difficult to measure than the amount sorbed. In most cases, sorption and desorption isotherms are nonsingular (i.e. for a given equilibrium solution concentration, more pesticide is held on the soil during the desorption phase than the sorption phase) or at times irreversible

Table IV. The measured vs predicted K_r values of sulfometuron methyl, terbuthiuon, and hexazinone on three soils as a function of soil depth if K_{oc} of the A horizon is assumed to be constant for other horizons. Predicted K_r values = [K_{oc} (A horizon) * OC for the horizon of interest]/100.

Soil*	Depth (cm)	OC (%)	Sulfometuron methyl Koc	Kd Measured	Predicted	Tebuthiuron Koc	Kd Measured	Predicted	Hexazinone Koc	Kd Measured	Predicted
1	0-5	4.27	13.10	0.56		84.1	3.59		25.70	1.10	
	5-10	1.02	12.60	0.13	0.130	54.4	0.55	0.860	19.50	0.20	0.260
	15-20	0.76	5.90	0.05	0.100	20.4	0.16	0.640	8.28	0.06	0.190
	65-70	0.07	45.00	0.03	0.010	59.6	0.04	0.060	26.90	0.02	0.020
	90-95	0.06	64.60	0.04	0.007	48.1	0.27	0.050	20.90	0.01	0.020
2	0-5	3.19	3.54	0.11		49.5	1.58		25.10	0.80	
	5-10	1.83	3.21	0.06	0.060	31.8	0.58	0.910	11.90	0.22	0.460
	15-20	0.76	5.91	0.05	0.030	12.5	0.09	0.380	6.20	0.05	0.190
	65-70	0.12	34.70	0.04	0.004	22.8	0.03	0.060	19.90	0.02	0.030
	90-95	0.10	27.90	0.03	0.003	32.5	0.03	0.050	34.30	0.03	0.030
3	0-5	8.25	1.25	0.10		24.6	2.03		9.42	0.78	
	5-10	1.61	4.36	0.07	0.020	32.5	0.52	0.390	12.40	0.20	0.150
	15-20	0.87	4.70	0.04	0.010	26.2	0.23	0.210	8.09	0.07	0.080
	65-70	0.04	96.50	0.04	0.001	83.9	0.03	0.009	72.20	0.03	0.004
	90-95	0.04	64.10	0.03	0.001	83.4	0.03	0.009	40.50	0.02	0.004

Modified with permission from reference 33, Chemosphere, copyright 1996.
*Soils correspond to soils described in reference 34.

(15). Therefore, prediction of the amount desorbed without actual measurement is nearly impossible. Clay and Koskinen (35) reported that between 9 and 71% of the hysteresis found between atrazine and alachlor sorption and desorption isotherms could be accounted for if the nondesorbable portion of chemical was determined.

Desorption has been reported to both increase and decrease with depth. For example, about 22% of sorbed alachlor in the A horizon of a silt loam soil was released during 5 consecutive desorption steps (36). However, no alachlor was desorbed from capillary fringe and saturated zone sediments (from 5.2 to 8 m) although the OC content was about 60-fold greater and the Kf was 13-fold greater in the A horizon. Desorption of imidacloprid showed the opposite trend, where more was desorbed from soil at lower depths compared to the amount desorbed from surface soil. This is indicated by the slope of the desorption isotherm ($1/n_{des}$) being more similar to the slope of the sorption isotherm ($1/n_{ads}$) lower in the profile (Table III).

Characterization of Soil Variability Effects on Sorption

Surface soil physical and chemical properties vary both spatially and temporally over field landscapes, affecting both sorption and desorption properties. Rarely are soil properties uniform from the soil surface to groundwater interface. Even within the aquifer zone, changes in soil texture, pH, and OC content can occur. In addition, soil properties vary through the soil profile even within a landscape making prediction of pesticide leaching a more difficult task. Data that examine spatial variation within pesticide sorption coefficients have been limited until recently.

Geostatistical analysis provides an approach for the characterization of spatially variable data (37, 38). For aggregated data, the variation between points (the semivariance statistic) is expected to increase as the lag interval (distance between points) increases out to some distance where spatial dependence is not detectable. A semivariogram is created to determine if there is spatial dependence in the data set. The x-axis of a semivariogram is the distance between sample pairs and the y-axis is the variation between points for all possible sample pairs at each distance. If semivariograms are positive and definitive, then there is spatial dependence between samples and unknown values between sample points can be estimated by kriging using weighting factors assigned for each sample point based on the spatial relationship.

The nugget, C_o, is interpreted as variability due to the combination of 1) experimental error, 2) distributional effects below the sampling scale of the study, and 3) other random effects. The value of the semivariance at the 1st lag distance, termed the experimental nugget, often is used as a more conservative estimate of the proportion of variability due to spatial structure (38).

The range, A, is the lag distance beyond which samples are considered spatially independent. The semivariance value at A is the maximum semivariance and is referred to as the sill (C_s). C_s represents the combination of the C_o effect and variability attributable to spatial dependence or structural variance (C), which is high frequency variance and is a process independent of spatial coordinates and only a function of lag distance. The proportion of variability attributable to spatial dependence may be estimated as % variability = $C/(C_o+C)$, where C_o is the nugget and C is variability caused by spatial dependence as indicated by the distance from C_o to C_s (maximum variance) (i.e. $C = C_s - C_o$) along the y-axis.

Another contribution to the total variance is drift, which is the total of large-scale low frequency variations in the data set (39, 40). Drift can be quantified once the nugget and sill are obtained from the fitted semivariogram model using the equation C_d = total variance - C_o – C, where C_d is the variability due to drift, C_o is the nugget, and C is the variability due to spatial dependence. Drift indicates that the prior mean of a process is a function of the spatial coordinates and the lag distance.

Spatial dependence of pesticide sorption coefficients (both Kd and Kf values) within the A soil horizon (or the 0 to 15 cm depth) across a field have been reported (3, 41, 42, 43). The sorption coefficients have been correlated to organic carbon (OC), sand and clay content (42, 44) and other properties such as pH (3, 33).

There are very few reports in the literature about spatial dependence of pesticide sorption in the vertical direction, i.e. from the soil surface to deeper depths in the soil (45, 46). The lack of this type of information is understandable because of the difficulty in obtaining soil samples at deep depths, and the number and volume of samples needed to estimate spatial variability. Jacques et al. (45) reported on the vertical spatial variability of atrazine sorption parameters in a loam soil. The CEC and OC content were found to have spatial structure with depth at lag distances of 0.5 and 1.5 m, respectively. The total variation when using a spherical model was made up of C_o, accounting for about 6% of the variance, C, accounting for about 18% of the variance, and C_d, accounting for 76% of the variance. Sand, silt, clay, and Kf had no spatial structure, i.e. the semivariogram showed pure nugget effect. When total variance was broken into C_o and C_d, C_o accounted for about 5% of the variance for silt and 15% for Kf. The C_d portion of the variance, the variability due to large-scale low frequency variations, accounted for 95% of the variance for silt and 85% of the variance for Kf. Jacques et al. (45) concluded that the estimation of Kf from basic soil properties (such as texture, OC, and CEC) should be restricted to the determination of large-scale variability (e.g. soil layers or different soil types).

In the study described above (45), it was not possible to get a good measurement of small-scale heterogeneity of Kf using regression equations based on measured soil properties. The correlation between two variables over spatial increments changed with spatial increment. For example, OC and Kf were highly correlated at vertical lag distances of 0.84 and 1.36 m but there was very little correlation at 0.18, 0.28, and 2.0 m.

Horizontal sampling at specified depths has also been used to evaluate spatial variability of chemical concentrations, specifically nitrate (47). Spatial correlation along the horizontal soil sample was highly significant at the three depths studied (41, 81, and 122 cm). The information gained about nitrate concentration at each depth from one horizontal core was estimated to replace between 7 to 21 vertical independent cores. The horizontal coring technique has other advantages including giving a more representative sample at the desired depth and not creating additional preferential flow paths from the surface.

Conclusions

There are several factors that influence pesticide adsorption to and desorption from soils. The sorption coefficient can be used in models to predict the potential of the pesticide to move to off-site locations through transport by such mechanisms as runoff or leaching. The spatial variations of pesticide sorption in the horizontal or vertical direction across field landscapes are important parameters in these predictions. Unfortunately, these relationships are difficult to obtain and in some cases the deterministic trend may be the overriding factor in the relationship. Depending on Koc values from surface soils to estimate sorption at lower depths can lead to erroneous results because, in most cases, sorption is underestimated by this technique. Either the quality of the OC at lower depths differs from that found at the surface or other mechanisms of sorption become more important.

While sorption and desorption parameters are important in modeling off-site movement, variability in other factors may influence pesticide transport. For example, deterministic models for pesticide transport were sensitive to Koc variability (6, 8). However, stochastic models used for actual field situations were dominated by large variability in soil parameters that control soil-water flow including hydraulic conductivity (8).

Indexes for sorption such as Kf, Kd, and Koc and biodegradation are useful to classify pesticides into relative mobility classes (48). It must be recognized, however, that a single index will never provide the data needed to quantitatively characterize pesticide transport in soil. Understanding the variability in these measurements across depths and landscape positions is extremely important to scientists to help predict probabilities for aquifer or surface water contamination or both at field, watershed, and regional levels. This knowledge can then be used by regulators to set policy that limits pesticide contamination in the environment and, ultimately, will help growers keep pesticides in their multifaceted tool box for pest control.

Acknowledgements

South Dakota Agricultural Experiment Station Manuscript No. 3205.

Literature Cited

1) Chancellor, W.J.; Goronea, M.A. *Trans. Am. Soc. Agric. Eng.*, **1994**, *37*, 717-724.
2) Boesten, J.J.T.I. *Pestic. Sci.* **1991**,*31* , 375-388.
3) Oliveira, R.S., Jr.; Koskinen, W.C.; Ferreira, F.A.; Khakurah, B.R.; Mulla, D.J.; Robert, P.J. *Weed Sci.* **1999**,*47* , 243-248.
4) Wagenet, R.J.; Rao, P.C.S. *Weed Sci.* **1985**, Suppl. 2, 25-32.
5) Brown,C.D.; Baer, U.; Gunther, P.; Trevisan, M.; Walker, A. *Pest. Sci.* **1996**, *47*, 249-258.
6) van der Zee, S.E.A.T.M.; Boesten, J.J.T.I. *Water Resour. Res.,* **1991**, **27**, 3051-3063.
7) Elabd, H.; Jury, W.A.; Cliath, M.M. *Environ Sci. Technol.,* **1986**, *20*, 256-260.
8) Lafrance, P.; Banton, O. *Geoderma,* **1995**,*65* , 331-338.
9) Koskinen, W.C.; O'Conner, G.A.; Cheng, H.H. *Soil Soc. Am. J.* **1979**, *43*, 871-874.
10) Calvet, R. *Interactions between herbicides and the soil;* Hance, R.J., ed.; Acad. Press: NY, pp.1-30.
11) Wauchope, R.D.; Meyers, R.S. *J. Environ. Qual.* **1982**, *30*, 14-20.
12) Schiavon, M. *Ecotox. Environ. Safety* **1988**, *15*, 55-61.
13) Khan, S.U. *Residue Rev.* **1982**, *84*, 1-25.
14) Green, R.E. *Pesticides in soil and water*; Guenzi, W.D.; Ahlrichs, J.L.; Bloodworth, M.E.; Chesters, G.; Nash, R.G., eds. Soil Sci. Soc. Am.: Madison, WI, **1974**, pp. 3-37.
15) Sparks, D.L. *Kinetics of soil chemical processes*; Acad. Press: San Diego, CA, **1988**, pp. 128-145.
16) Laird, D.A.; Bariusso, E.; Dowdy, R.H.; Koskinen, W.C. *Soil Sci. Soc. Am. J.,* **1992**,*56* , 62-67.
17) Ukrainczyk, L.; Ajwa, H.A. *Soil Sci. Soc. Am. J.,***19 96**, *60*, 460-467.

18) Celis, R.; Cornejo, J.; Hermosin, M.C.; Koskinen, W.C. *Soil Sci. Soc. Am. J.*,**19 98**,*62* , 165-171.

19) Johnson, R.M.; Sims, J.T. *Pestic. Sci.*,**19 98**,*54* , 91-98.

20) Brady, N.C.; Weir. R.R. *Elements of the nature and properties of soils*. Prentice Hall, Upper Saddle River, NJ, **2000**, pp. 59-94.

21) Karickhoff, S.W. *Chemosphere*, **1981**,*10* , 833-846.

22) Di, H.J.; Aylmore, L.A.G., *Soil Sci. Soc. Am. J.*,**19 97**, *61*, 17-23.

23) Pang, L.P.; Close, M.E.; Watt, J.P.C.; Vincent, K.W. *J. Contamin. Hydrol.*, **2000**, *44*, 16-46.

24) Seybold, C.A.; McSweeney, K.; Lowery, B. *J. Environ. Qual.*, **1994**, *23*, 1291-1297.

25) Norman, R.J.; Kurtz, L.T.; Stevenson, F.J. *Soil Sci. Soc. Am. J.*, **1987**, *51*, 809-812.

26) Kissel, D.E.; Cabrera, M.L.; Ferguson, R.B. *Soil Sci. Soc. Am. J.*, **1988**, *52*, 1753-1796.

27) Clay, S.A.; Koskinen, W.C.; Allmaras, R.R.; Dowdy, R.H. *J. Environ Sci. Health,* **1988**,*B23* , 559-573.

28) Stougaard, R.N.; Shea, P.J.; Martin, A.R. *Weed Sci.*,**19 90**, *38*, 67-73.

29) Liu, Z.; Clay, S.A.; Clay, D.E.; Harper, S.S. *J. Agric. Food Chem.* **1995**, *43*, 815-819.

30) Clay, S.A.; Clay, D.E.; Liu, Z.; Harper, S.S. *Herbicide metabolites in surface and groundwater;* Meyer, M.T.; Thurman, E.M., eds.;ACS Symposium Series 630. Amer. Chem. Soc. Clarendon Hills, IL; **1996**, pp. 117-124.

31) Liu, Z.; Clay, S.A.; Gaffney, J.; Malo, D. *South Dakota Acad. Sci.* **1997**, *76*, 153-164.

32) Felding, G. *Pestic. Sci.*, **1997**,*50* , 64-66.

33) Koskinen, W.C.; Stone, D.M.; Harris, A.R. *Chemosphere*, **1996**, *32*, 1681-1689.

34) Harris, A.R.; Stone, D.M. *Research paper NC-291,* **1990**, USDA-Forest Service, St. Paul, MN.

35) Clay, S.A.; Koskinen, W.C. *Weed Sci.*, **1990**, *38*, 262-266.

36) Clay, S.A.; Moorman, T.B.; Clay, D.E.; Scholes, K.A. *J. Environ. Qual.* **1997**, *26*,1348-1353.

37) Rossi, R.E.; Mulla, D.J.; Journel, A.G.; Franz, E.H. *Ecol. Monogr.* **1992**, *62*, 277-314.

38) Williams, L. III; Schotzko, D.J.; McCaffrey, J.P. *Environ. Entomol.* **1992**, *21*, 983-995.

39) Russo, D.; Jury, W.A. *Water Resour. Res.*, **1987**,*23* , 1269-1279.

40) Russo, D.; Bouton, M. *Water Resour. Res.*,**19 92**, *28*, 1911-1925.

41) Rao, P.C.S.; Edvardsson, K.S.V.; Ou, L.T.; Jessup, R.E.; Nkedi-Kizza, P.; Hornsby, A.G. *Evaluation of pesticides in groundwater;* ASC Symposium Series 315, Amer. Chem. Soc. Clarendon Hills, IL; **1986**,p p. 100-115.

42) Liu, Z.; Clay, S.A.; Clay, D.E. *In Proceedings 4th International Conference Precision Ag.;* Robert, P.J.; Rust, R.H.; Larson, W.E., eds. Soil Soc. Amer. Madison, WI; **1998**, pp. 879-883.

43) Novak, J.M.; Moorman, T.B.; Cambardella, C.A. *J. Environ. Qual.* **1997**, *26*, 1271-1277.

44) Clay, S.A.; Dowdy, R.H.; Lamb, J.A.; Anderson, J.L.; Lowery, B.; Knighton, R.E.; Clay, D.E. *J. Environ. Health Sci.* **2000**, *B35*, 259-278.

45) Jacques, D.; Mouvet, C.; Mohanty, B.; Vereecken, H.; Feyen, J. *J. Contamin. Hydrol.* **1999**, 36, 31-52.

46) Beck, A.J.; Harris, G.L.; Howse, K.R.; Johnson, A.E.; Jones, K.E. *Chemosphere*, **1996**,*33* , 1283-1295.

47) Verhoeff, R.L.; Powers, W.L.; Shea, P.J.; Marx, D.B.; Wieman. *J. Soil Water Conserv.*,**19 97**,*52* , 126-131.

48) Jury, W.A.; Focht, D.D.; Farmer, W.J. *J. Environ. Qual,* **1987**, *16*, 422-428.

Chapter 9

In Situ Measurements of Soil Hydrology in a Field Dissipation Study: Implications for Understanding Pesticide Residue Movement

Wenlin Chen[1], Thomas Wiepke[1], and Nathan Snyder[2]

[1]Health Assessment and Environmental Safety, Syngenta, 410 Swing Road, Greensboro, NC 27409
[2]Waterborne Environmental, Inc., 897B Harrison Street, Leesburg, VA 20175

Data of pesticide soil residue from terrestrial field dissipation studies reflect dissipation of all routes available under field conditions. Separate evaluation of key dissipation routes is often difficult but essential to a better understanding and assessment of the integrated environmental fate processes. This paper presents some preliminary results of a field dissipation experiment designed not only to evaluate the dissipation of a test compound, but also to understand the in-situ field hydrology via intensively monitored soil water by Time Domain Reflectometry (TDR), and the use of KBr as a conservative tracer. With the measured soil moisture data, daily water balance including groundwater recharge and potential runoff is evaluated. An integrated assessment of the environmental fate and transport processes based on field residue data is also provided.

Introduction

Chemical fate and transport processes in soil are influenced by soil hydrological factors including the dynamic soil water regime and hydraulic properties. While soil water carries dissolved chemicals moving through soil pores downwards or upwards depending on the gradient direction of the soil water potential, the transport of chemicals in field soils are often complicated by transient surface boundary conditions and soil spatial heterogeneity. Interpreting field data thus can be difficult without the corresponding perspectives of soil hydrology.

The current US Environmental Protection Agency (EPA) terrestrial field dissipation study guidelines emphasize the objective to identify the overall dissipation pattern of a test compound under field use conditions and to assess the mobility potential. However, current groundwater risk assessment and modeling analysis rarely use field dissipation data primarily due to the concern that potential leaching or runoff may contribute to the overall field dissipation. Most field dissipation studies do not measure field soil hydrological processes. Consequently, separate evaluation of potential field leaching and runoff is often difficult. This paper presents some preliminary results of a field dissipation experiment designed not only to evaluate the dissipation of a test compound, but also to understand the in-situ field hydrology via intensively monitored soil water by Time Domain Reflectrometry (TDR), and the use of a conservative tracer (KBr). The analysis focuses on how field hydrological measurements can be employed for an integrated assessment of environmental fate and transport behavior of a test compound. Water movement data along with the test compound residue concentrations are interpreted by the numerical model, LEACHP (*1*).

Materials and Methods

A site in the coastal plain of North Carolina (Sampson County), USA, was selected for the field dissipation experiment. The field has a minimum slope (<1%) and was not cropped during the experimental period (i.e., a bare soil plot). The site-specific soil was a Norfolk loamy sand (2) with the soil chemical-physical properties listed in Table 1. The soil bulk density and water holding capacity (WHC) at both 1/3 and 15 bar were measured using undisturbed soil cores taken at five randomly selected locations in the test plot (27 m x 30 m). The test plot layout and instrumentation location for measurement of soil moisture content (TDR), temperature, and rainfall, along with Guelph permeameter measurement locations are illustrated in Fig. 1.

Field Soil Saturated Hydraulic Conductivity

Field soil hydraulic conductivity K_{fs} in the vadose zone was measured using a Guelph permeameter (GP) system (Soilmoisture Model No. 2800K1) (3) at the beginning of the study. The Guelph constant-head permeameter measures the rate of infiltration of water into a borehole while maintaining a constant head within such borehole. The field measured value of K_{fs} may differ from laboratory saturated hydraulic conductivity (K_{sat}) in that K_{fs} may account for entrapment of air bubbles within the soil structure, thus it is likely to provide a more realistic value for soils in the unsaturated zone.

Table 1. Site-Specific Average Soil Chemical-Physical Properties

Soil Depth (cm)	USDA Soil Texture	BD (g/cm³)	Moisture Content at 1/3-Bar (cm³/cm³)	CEC (meq./100g)	Soil pH	OM (%)
0 – 15	LS	1.57	0.12	4.7	6.5	1.4
15 – 30	LS	1.71	0.12	4.1	6.1	1.0
30 - 45	SCL	1.53	0.22	7.3	5.8	0.7
45 - 60	SCL	1.56	0.32	7.0	5.4	0.3
60 - 75	SCL	1.55	0.34	8.0	5.2	0.2
75 - 90	SCL	1.56	0.36	7.2	5.1	0.1
90 -105	SC	1.58	0.37	7.4	5.0	0.1

NOTE: LS=Loamy Sand; SCL=Sandy Clay Loam; SC=Sandy Clay. BD=Bulk density; OM=Organic Matter Content; CEC=Cation Exchange Capacity. Bulk density and 1/3 bar moisture content were measured on undisturbed soil cores taken from the test plot.

Four locations within and along the test plot borders (27 m x 30 m, Fig. 1) were chosen to examine the spatial variability of K_{fs}. Measurements were completed at three depths (15, 40, and 107 cm, coinciding with the midpoint of each natural soil horizon) and data was collected for at least two different heads at each depth. The rate of infiltration, head of water, and other data pertaining to the size of the reservoir and borehole were recorded once a steady-state infiltration rate was achieved for each measurement. Upon completion of each test, native soil was used to refill the hole to the ground surface.

Three methods were involved in the analysis of the GP data to calculate K_{fs}. These were the Laplace/Gardner method (LGM) (4), the one-ponded height method (OPHM) (5), and the simultaneous equation method (SEM) (4). As documented in the literature, each calculation approach has its merits and limitations. The analysis adopted in this study was to take the average results of all the three methods for each measurement as shown in Table 2.

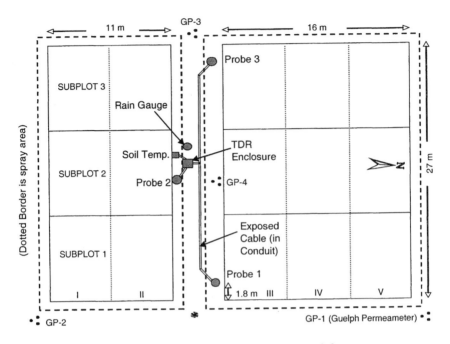

Figure 1. Field plot layout, instrumentation and Guelph permeameter measurement locations.

Table 2 . Test Site Field-Saturated Hydraulic Conductivity Measured by the Guelph Constant-Head Permeameter

Location	Depth 15 cm	Depth 40 cm	Depth 107 cm
GP-1	2.54	0.51	0.102
GP-2	1.78	1.78	0.254
GP-3	30.48	0.76	0.254
GP-4	12.70	0.51	0.102
Field Average	11.87	0.89	0.178

NOTE: Units are cm/hr. Results of each location were averages of three calculation methods (i.e., the Laplace/Gardner method, the one-ponded height method, and the simultaneous equation method).

Soil Moisture and Meteorological Measurements

A Time Domain Reflectometry (TDR) system was installed to measure hourly soil moisture content (volumetric) at three locations, one in each subplot (Fig. 1). Measurements were made to a depth of 105 cm in continuous 15 cm increments from the soil surface. An Environmental Sensors Incorporated (ESI) Moisture Point™ TDR system was used. Measurements were transmitted to and stored by a Campbell Scientific Inc. (CSI) CR10X data logger. The datalogger controlled the multiplexer which switched between the probes. The datalogger was also used to record soil temperature which was measured by four CSI 107B soil temperature thermistor probes (data not presented). Accessory equipment included the support structure, instrument enclosure, solar recharged battery power supply, and data storage modules (CSI SM192). Two storage modules were used alternately to store and download collected data between the TDR system and a PC computer.

Table 3. Comparison of TDR and Soil Core Measured Soil Moisture Contents on June 9, 1999 (Unit: cm^3/cm^3)

Soil Depth	TDR (east)	Soil Core (east)	TDR (center)	Soil Core (center)	TDR (west)	Soil Core (west)
0-15 cm	0.121	0.133	0.103	0.136	0.112	0.132
15-30 cm	0.126	0.144	0.106	0.128	0.118	0.130
30-45 cm	0.269	0.218	0.309	0.251	0.198	0.162
45-60 cm	0.294	0.256	0.272	0.267	0.26	0.218
60-75 cm	0.26	0.272	0.313	0.285	0.318	0.215
75-90 cm	0.28	0.275	0.337	0.293	0.331	0.225
90-105 cm	0.35	0.289	0.389	0.293	0.383	0.249

Three TDR probes with seven 15-cm soil depth interval measurements were installed from the surface in an alley adjacent to each subplot of the dissipation study with the measurement equipment (TDR enclosure) located at the central (Fig. 1). Each probe had seven measurement sections for seven 15-cm soil depth intervals. One probe was installed near the central location (subplot 2) and the other probes were installed adjacent to subplot 1 and subplot 3. The probes were installed in an alley to avoid interference with the tractor traffic for the application and maintenance activities of the dissipation study.

Soil cores were collected at one time point for moisture determination in order to provide some verification of the TDR measurements. Table 3 compares the TDR readings with the soil core measured moisture contents at the three TDR installation sites. Note that since the TDR measurements were on the volumeric basis, the gravimetrically determined soil moisture contents were converted to the volumetric convention in Table 3 using the undisturbed soil bulk density at each corresponding depth. As shown in Table 3, the TDR readings agreed generally well with the soil core data even though there was some observable moisture loss from the soil core samples, particularly at the lower depth, during transport from the field to laboratory, i.e., condensation on the plastic liner for the soil cores was observed.

Daily meteorological data including precipitation, air temperature, and wind speed were measured from an on-site weather station (about 300 m away from the west side of the field). Pan evaporation data were obtained from NOAA Hofmann Forest weather station (about 60 km from the study site). Irrigation using a lateral moving overhead system was only applied as needed to achieve 120% of the NOAA precipitation norm for a month. Rain gauges were placed in the plot to measure the amount of irrigation water delivered for each event during the experimental period (June 8, 1999 - March 25, 2000).

Tracer/Test Substance Application and Sampling

Two separate broadcast applications of KBr (112 kg/ha) and a test substance (0.336 kg/ha) were made to the bare soil field plot on June 9, 1999 using a tractor-mounted boom sprayer. Soil cores were collected from the treated and control plots prior to application, immediately after application, and then at specified intervals through March 25, 2000. At each sampling event, five cores were collected from each of the three treated subplots for a total of 15 cores. The soil cores were taken to a depth of 120 cm and divided into 15 cm increments. Each set of five cores was composited by depth increment, resulting in three replicate samples (A, B, and C) per depth increment for laboratory analysis. Bromide anion concentration in samples was determined using ion chromatography (0.20 ppm, Limit of Quantification or LOQ). The test substance concentration in samples was determined using HPLC with MS detection (5.0 ppb LOQ).

Modeling

The integrative nature of the field data involving transient conditions of simultaneous meteorological, hydrological, and environmental fate processes requires a comprehensive numerical model to aid the data analysis. LEACHP, a pesticide fate module of LEACHM (1), was used for this purpose. LEACHP uses the Richards equation to simulate the transient soil water flow primarily driven by water potential gradients in the vadose zone:

$$\frac{\partial \theta}{dt} = \frac{\partial}{\partial x}\left[K_h(\theta)\frac{\partial H}{\partial x} \right] - U(x,t) \tag{1}$$

where θ is volumetric water content in soil ($m^3 \cdot m^{-3}$); t is time (day); x is distance from soil surface (mm); K_h is soil hydraulic conductivity ($mm \cdot day^{-1}$); H is hydraulic head in unsaturated soil (sum of soil matrix and gravitational potential, mm); U is a sink term representing water loss by transpiration (or root uptake rate, day^{-1}).

The upper boundary conditions were assumed as $H=0$ at $x=0$ during ponded infiltration, or otherwise:

$$-K_h(\theta)\frac{\partial H}{\partial x} = q_0(t) \qquad \text{at } x=0$$

where q_0 is either infiltration rate estimated based on rain/irrigation events, or soil evaporation (i.e., $-E_s$, $mm \cdot day^{-1}$) calculated from pan data as mentioned previously. The lower boundary was assumed as a free drainage profile, or mathematically, $\partial H/\partial x=1$ at $x = 3$ m.

Parameterization of the soil hydraulic properties required by eq 1 was provided by LEACHP based on the measured basic soil physical-chemical properties (Table 1, 2, and 4). Specifically, the soil-water retention function was estimated using the Rawls and Brakensiek regression based on soil particle size distribution, organic matter content, and bulk density (1). The Campbell (1) equation was used for K_h as a function of soil moisture contents.

The convection-dispersion equation (CDE) was used to describe the chemical transport process coupled with chemical soil degradation, and sorption kinetics in LEACHP (1) (gas phase was not considered here):

$$\frac{\partial(\theta + f\rho_b K_d)C}{dt} = \frac{\partial}{\partial x}\left(\theta D\frac{\partial C}{\partial x} - qC \right) - \rho_b\frac{\partial S_k}{\partial t} - \theta k_{-1}C \tag{2}$$

$$\frac{\partial S_k}{dt} = k_{des}\left[(1-f)K_d C - S_k\right] \tag{3}$$

where C is concentration in soil pore water ($mg \cdot L^{-1}$); q is water flux in soil ($mm^3 \cdot day^{-1}$); S_k is concentration on the kinetic sorption site ($mg \cdot kg^{-1}$); f is the fraction of the equilibrium sorption site (the mass fraction of soil comprising equilibrium sorption site); ρ_b is soil bulk density (g·cm^{-3}); θ is soil volumetric water content (cm^3·cm^{-3}); K_d is adsorption equilibrium partition coefficient between the soil solid and aqueous phases (ml/g); k_{des} is desorption rate constant (day^{-1}); k_{-1} is degradation rate constant in soil pore water constant (day^{-1}); D is the effective soil aqueous phase diffusion/dispersion coefficient (mm^2·day^{-1}) which is the sum of molecular diffusion and mechanical dispersion in soil. A dispersivity 30 mm was used in this study and the molecular diffusion coefficient in soil pore water was estimated by LEACHP. For both bromide and the test compound, the initial soil concentrations (prior experiment) were zero. No chemical flux out of the soil surface was allowed during evaporation and a constant concentration boundary condition was assumed at the bottom of the simulated soil profile.

Results and Discussions

Soil Spatial Variability in Test Plot

The soil hydraulic properties measured at five randomly selected locations in the experimental plot provided some indication of the soil spatial variability (Table 4). Among the measured parameters, the soil WHC at both 1/3 and 15 bar appeared more variable than soil bulk density relative to their mean values.

Table 4. Spatial Variation of Soil Bulk Density, 1/3 and 15 Bar Moisture Retention Content

Depth	Mean of 5 sample locations			Standard Deviation		
(cm)	Bulk density	1/3 bar	15 bar	Bulk density	1/3 bar	15 bar
0 – 15	1.57	0.12	0.05	0.043	0.0045	0.0026
15 – 30	1.71	0.12	0.06	0.073	0.0060	0.0113
30 - 45	1.53	0.22	0.18	0.087	0.0818	0.0792
45 - 60	1.56	0.32	0.27	0.088	0.0484	0.0544
60 - 75	1.55	0.34	0.29	0.042	0.0321	0.0342
75 - 90	1.56	0.36	0.32	0.032	0.0432	0.0440
90 -105	1.58	0.37	0.32	0.034	0.0457	0.0397

NOTE: 1/3 and 15 bar are volumetric soil water contents at 1/3 and 15 bar pressure, respectively. Unit of soil bulk density, g/cm^3; moisture content, cm^3/ cm^3.

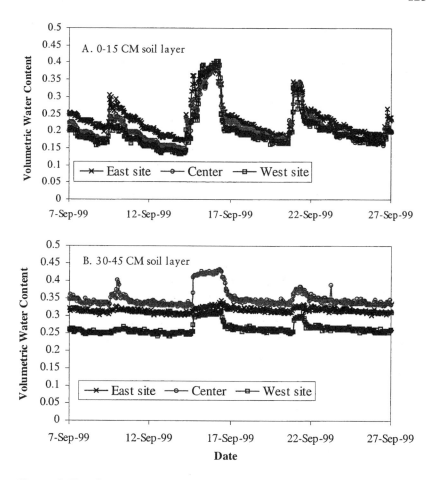

Figure 2. Hourly TDR measured soil moisture content at three different field locations during September 7 to 27, 1999.

Compared to the bulk density and the two water retention parameters at 1/3 and 15 bar, the in-situ measured field soil saturated hydraulic conductivity (K_{fs}) was found to be more variable in the top soil (0-30 cm) corresponding to the plow layer (Ap) (Table 2). The measured K_{fs} value ranged from 1.8 to 30 cm/hr in this layer, with the lower values increasing from the eastern side to the west of the field plot (Fig. 1). This spatial trend, however, was not seen for values measured at two lower soil depths (Table 2). The spatial variability in the top soil layer may reflect the past farming practices. The test area has been in a rotation of tobacco, corn, fallow, cucumber, and tobacco each year since 1994. Prior to the experiment, the soil was chisel-plowed (approximately 30-36 cm deep), disked (about 10-15 cm deep), and then firmed with a small tractor with turf tires after the cultivation.

From soil borings extended at the test site, there appeared to be some heterogeneity between soil horizons as one moves from the eastern to the western borders of the test plot. The borings extended on the eastern border and middle of the test plot (locations GP-1 , GP-2 and GP-4) were similar to well-drained Norfolk loamy sand (2), with a sand/loamy sand Ap horizon and a B horizon of sandy clay loam having low organic matter contents and clay contents from approximately 26 to 38 percent. Whereas the soil horizon encountered at the western border of the test plot, GP-3, had a deeper A horizon and much higher organic matter content with depth in a transitional A/B horizon extending to 74 and 89 cm. These differences were also reflected during the installation process of the TDR probes in which the western probe was significantly easier to install then the central and western probe.

The effect of spatial variability in soil hydraulic conductivity is reflected in the TDR measurements (Fig. 2). The TDR-measured soil water content at the three different field locations was observably different over the same time period from September 7 to 27, 1999. Soil water content was held almost constantly higher at the eastern location (Fig. 2A), i.e., less drainage because of much lower permeability in the top soil layer than at the western location (Table 2). On the other hand, arrival of peak water content corresponding to rainfall or irrigation at the lower depth of 30-45 cm was delayed or diminished at the eastern location due to the lower water hydraulic conductivity on the top.

Soil Water Dynamics

Despite the potential soil spatial variability as discussed above, the TDR measured soil moisture data at three field locations were averaged for each soil depth (every 15 cm segment). The average TDR results were then compared to the LEACHP predicted values using the average site-specific soil hydraulic input parameters (Table 1 and 2) (Fig. 3). As illustrated in Fig. 3, the dynamic patterns (peaks and valleys) in both the measured and simulated results were more significant in the top soil layers than in the lower layers, indicating more active processes of evaporation, irrigation/precipitation, and drainage in the top soil horizon.

LEACHP predictions of the soil moisture dynamic pattern appeared to agree reasonably well with the measured data particularly in the active top soil layer (Fig. 3A). Although the model tended to overpredict the water content of the 30-

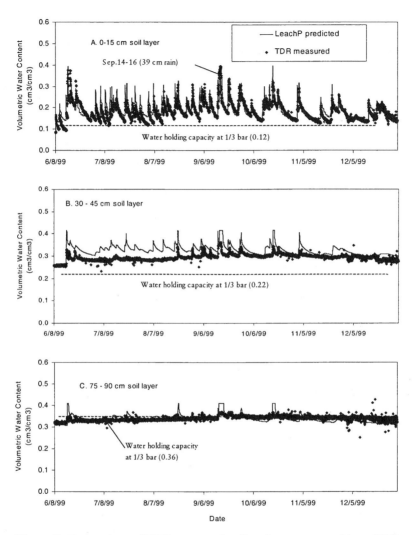

Figure 3. Comparison of TDR measured soil moisture content (three TDR location average) to LEACHP predicted for the soil layer 0-15 cm (A), 30-45 cm (C), and 75-90 cm (B) during June 8, 1999 to January 4, 2000. Each solid diamond represents a TDR measurement at each hour.

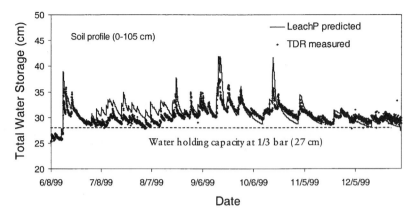

Figure 4. Hourly TDR measured total water storage in soil profile (0-105 cm, three TDR location average) compared to LEACHP predicted. Dashed line represents field soil water holding capacity measured at 1/3 bar pressure in laboratory.

45 cm soil layer, the overall prediction of the water storage in the entire soil profile (0-105 cm) was reasonable (Fig. 4). Compared to the measured data, the laboratory measured field WHC at 1/3 bar was generally well below the field moisture content in the top soil layers (0-15 and 30-45 cm), about 30-40% less than the average soil water content held during the entire experimental period. The field WHC at depth 75-90 cm (Fig. 4C) or below (not shown) appeared to be close to the average water content probably due to the finer soil texture (Table 1), lower hydraulic conductivity (Table 2), and proximity to the ground water table.

The dynamic nature of field soil moisture at levels constantly above WHC indicates that the drainage process may be oversimplified by assuming soil water drains only when the WHC is exceeded (i.e. the tipping-bucket type of models). It also underlines the importance of WHC calibration based on field data when such a model is used. Additionally, the intensively measured TDR data demonstrated the continuous fluctuation of the soil moisture content with rise and fall corresponding to periodic changes of rain/irrigation and sunshine (evaporation) conditions on the field. The vigorous fluctuation and continuous redistribution of water in the entire soil profile, downward or upward as determined by the water potential gradient, supported the diffusion-based LEACHP model which adjusts time steps other than a fixed daily interval for accurate soil water calculation.

Water fluxes, the amount of water moved through various soil depths during a specific time period, were estimated by LEACHP as illustrated in Fig. 5. The model defines upward flux as negative while downward flux as positive. It is clearly demonstrated that downward movement was predominant during the entire experimental period due to the substantial amount of water input from rain and irrigation (total 138 cm). The highest drainage flux as shown in Fig. 5 occurred during the period of hurricane Floyd which deposited 39 cm of rain over three days (September 14-16). Water fluxes appeared much higher in the upper soil profile than at the bottom, indicating flexible water storageability by soil.

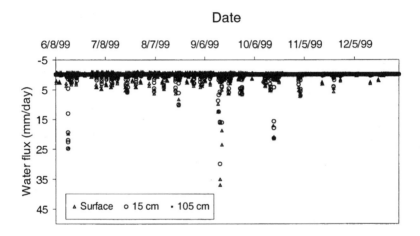

Figure 5. LEACHP simulated water fluxes across various soil depths (hourly output data). Positive values indicate downward flow while negative values indicate upward flow.

Soil Profile Water Balance

The cumulative water balance including water input (rain and irrigation), surface evaporation, and drainage at 105 cm is illustrated in Fig. 6. Since hourly measured soil moisture contents were available, hourly water balance was estimated based on the following equation:

$$\Delta = Water\ Input - ET - Storage - Drainage \qquad (4)$$

Given water input (precipitation and irrigation), storage, and evaporation that were obtained from measured data, the magnitude of water balance Δ in eq 4 would represent either error or potential runoff that was ignored in LEACHP. Plotted in Fig. 7 is the hourly water balance Δ calculated based on eq 4. Fluctuations of the calculated value around the perfect balance (i.e. zero) may reflect potential inaccuracies in estimating each of the water balance components in eq 4. Potential surface runoff, however, might not have been a significant factor contributing to the fluctuations since there were no consistent biases to the positive side of the equation even though the fluctuation tended to be magnified during large precipitation events (Fig. 7 and 8). The magnified fluctuation was probably due to the limited numerical accuracy of the LEACHP calculated drainage when soil water flux was high during high rain events. Generally, the fluctuation was small with its daily moving average approaching zero. This is particularly true when hourly precipitation data were used for LEACHP to estimate drainage (Fig. 8).

Table 5. Water Balance (6/8/99 - 12/31/99) in the Soil Profile 0-105 cm.

	Total Water Input	ET	Drainage	All Other
Amount (cm)	138.34	49.85	85.76	2.73
% Total Water Input	100%	36%	62%	2%

NOTE: Drainage was Estimated by LEACHP.

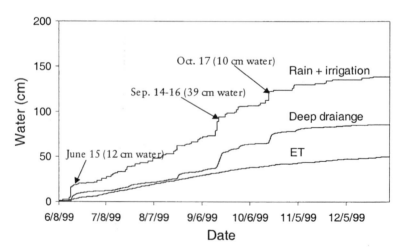

Figure 6. Cumulative water input and output in the soil profile 0-105 cm during June 8, 1999 to January 4, 2000.

The total water balance for the soil profile 0-105 cm during the period from June 8 to December 31, 1999 is listed in Table 5. As indicated in the table, 62 % of all received water drained below 105 cm, 32 % evaporated, and only 2 % might have contributed to soil profile storage change or potential surface runoff.

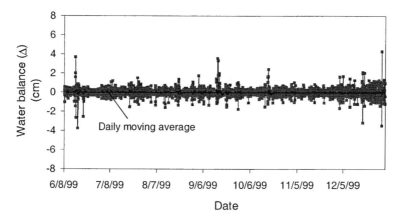

Figure 7. Soil profile water balance (0-105 cm) using daily precipitation/irrigation data (June 8, 1999 to January 4, 2000).

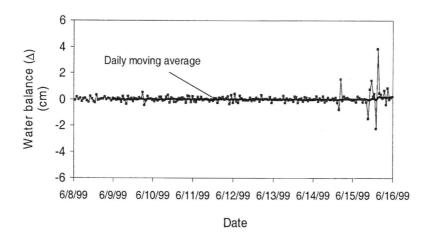

Figure 8. Soil profile water balance (0-105 cm) using hourly precipitation/irrigation data (June 8-16, 1999).

132

Chemical Transport

Bromide soil concentrations measured in the top soil layer (0-15 cm) are compared with the LEACHP predicted results in Fig. 9 over the first month after application on June 9, 1999. Bromide stayed in the top soil layer without much downward movement until June 14 when there was about 14 cm rain event that occurred June 14 - 15. As LEACHP simulated, concentrations of bromide dropped dramatically due to downward leaching after the two day rainfall event.

The distribution of bromide in the soil profile is better illustrated in a snapshot plot for June 30, about three weeks after application (Fig. 10). As seen in the figure, bromide moved much lower with the center of mass located at near 45 cm after the substantial rain during the week of June 15. Comparatively, the center of mass for the test substance was located near the 14 cm depth, indicating retardation of the pesticide due to soil adsorption. Bromide was fairly symmetrically distributed along the soil profile and was described reasonably well by LEACHP, suggesting that field preferential flow was not significant in the sampled area.

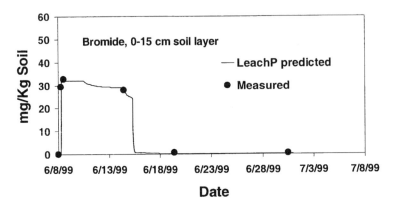

Figure 9. Comparison of measured Br⁻ soil concentrations (solid circles) to LEACHP predicted (line) for the soil layer 0-15 cm during June 8, 1999 to January 4, 2000.

As shown in Fig. 11, the residue concentrations of the test substance in the top soil layer (0-15 cm) dissipated rapidly, with <18% of the day 0 concentration remained in the soil five days after application. No residue was detected at or above the 5 ppb LOQ for soil layers below 15 cm in the field study. Compared to the bromide dissipation curve (Fig. 9) in the same soil layer, the test substance appeared to dissipate continuously with a gradual decline. The impact of the large rainfall events during June 14-15 on the substance dissipation, was not as evident as with the bromide, probably due to the mitigating effect of binding of the test substance to the soil matrix.

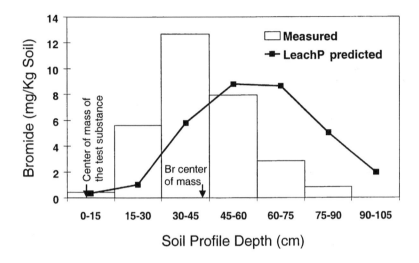

Figure 10. Snapshot of measured Br⁻ concentrations on June 30, 1999 in the soil profile 0-105 cm (bar) compared to the LEACHP predicted (line) on the same day. Arrows represent the center of mass of Br⁻ and the test substance observed in the soil profile.

LEACHP predictions of the test substance were carried out with two distinct approaches with regard to interactions between degradation and adsorption/ desorption kinetics (Fig. 11). The first approach assumes instantaneous adsorption/desorption and first-order degradation taking place at the same rate on both dissolved (pore water) and adsorbed phases in soil (referred to as the one-compartment model) (6). The second approach adopts the theory that degradation occurs only in pore water while time-dependent adsorption/ desorption controls the mass transfer process between the dissolved and adsorbed phases (referred to as the two-compartment model) (6). Time-dependent adsorption/desorption of pesticide by soil particles has been observed widely and are often attributed to a number of reasons including specific reactions of adsorption bonding kinetics, diffusive movement within intra-

134

organic matrices or immobile fluid in the intra-particle/intra-aggregate micropores in soil (7-9).

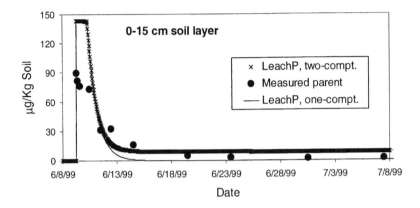

Figure 11. Measured parent residue soil concentrations (solid circles) compared to LEACHP predicted (lines) for the soil layer 0-15 cm during June 8 to July 8, 1999. The thiner line presents LEACHP simulations assuming one-compartment expontial degradation, whereas the thicker line assumes a two compartment degradation model in LEACHP.

Predictions by the two approaches as selected in LEACHP are compared with the measured residue data in Fig. 11. The environmental fate parameters of the test substance were determined from laboratory adsorption/desorption batch equilibrium and soil metabolism studies, including the soil organic carbon adsorption partition coefficient (Koc=372), desorption rate constant (α=0.00315 day^{-1}) and degradation half-life in dissolved phase ($T_{1/2}$=0.61 days). As illustrated in Fig. 11, both one- and two- compartment models predicted the peak residue concentrations reasonably well. The flat peak in both model predictions was due to the fact that the predicted peak concentrations exceeded the compound water solubility in the soil pore water (100% application efficiency was assumed in LEACHP). LEACHP assumes no degradation when a compound has precipitated from solution. Compared to the one-compartment model, the two-compartment model tended to simulate the residue tail better, indicating a rate limiting effect of the desorption kinetics on the compound availability to degrade. The aged residue, often tightly bound, has to slowly desorb to the soil solution before degrading.

Conclusions

The field soil water regime, both as extensively monitored and simulated, demonstrated significant downward water movement. Field water balance analysis indicated no significant surface runoff even under extreme water input events (e.g. 22 cm /day). During the experimental period between June 8 to December 31, 1999, a total of 138 cm water (precipitation and irrigation) was received by the test plot. 62 % of the total water input was estimated to have drained below 105 cm, 32 % evaporated, and only 2 % may have contributed to soil profile storage change or potential surface runoff. Water retention by field soil may be severely underestimated by laboratory measured soil water holding capacity. This observation, therefore, underlines the importance of model calibration against field soil water measurements particularly for "tipping-bucket" type models. The rapid dissipation and slow movement of the test substance contrasted with the bromide data, and the predominant downward water fluxes during the entire experimental period, indicates that the test substance is not likely to leach significantly once applied to soil.

Acknowledgment

Assistance from colleagues, P. Scott, P. Manuli, L. Swaim, K. Kabler, R. Speth, Drs. R. W. Williams, K. Winton, and S. Chen of Syngenta Crop Protection, Inc. is much appreciated.

References

1. Hutson, J.L.; Wagenet, R.J. LEACHM: Leaching Estimation and Chemistry Model. Version 3. Department of Soil, Crop and Atmospheric Sciences, Cornell University, Ithaca, New York 14853, USA. Research Series No. 92-3, September 1992.
2. Brandon, C.E. Soil Survey of Sampson County, North Carolina, United States Department of Agriculture, Soil Conservation Service, 1985.
3. Soilmoisture Equipment Corporation. 2800KI Guelph Permeameter Operation Instructions, Rev. 8/86, October 1991.
4. Reynolds, W.D.; Elrick, D.E. Ground Water Monitoring Review. **1986.** 6, 84.
5. Elrick, D.E.; Reynolds, W.D.; Tan, K.A. Ground Water Monitoring Review. **1989.** 9, 184.

6. Van Genuchten, M.Th; Wagenet RJ. Soil Sci. Soc. Am. J. **1989**. 53, 1303.
7. Chen, W.; Wagenet, R.J. Soil Sci. Soc. Am. J. **1997**. 61, 360.
8. Pignatello, J.J.; Ferrandino F.J., Huang, L.Q. Environ. Sci. Technol. **1993**. 27, 1563.
9. Ball, W.P.; Roberts, P.V. Environ. Sci. Technol. **1991**. 25, 1237.

Chapter 10

Electronic Soil Moisture Measurements in Federal Insecticide, Fungicide, and Rodenticide Act Field Dissipation and Prospective Groundwater Studies

Nathan J. Snyder[1], Juliet M. Cartron[2], Ian van Wesenbeeck[3], Les S. Carver[1], and Amy M. Ritter[1]

[1]Waterborne Environmental, Inc., 897B Harrison Street, Leesburg, VA 20175
[2]Greenhorne and O'Mara, Inc., 9001 Edmonston Road, Greenbelt, MD 20770
[3]Dow AgroSciences, Building 306, A2, 9330 Zionsville Road, Indianapolis, IN 46268–1053

The draft United States Environmental Protection Agency (USEPA) guidelines for Federal Insecticide, Fungicide, and Rodenticide Act (FIFRA) Terrestrial Field Dissipation and Prospective Ground-Water (PGW) studies have soil-moisture monitoring requirements. The measurement methodology, frequency, and locations are at the discretion of the registrant. Time Domain Reflectometry (TDR) was used to monitor soil moisture at two PGW study sites, conducted to fulfill part of the registration requirements for a soybean herbicide in the United States. Measurements of soil moisture were continuous from the soil surface to a 3.6-m depth, close to the water table, and recorded hourly. A bromide tracer was applied and transport monitored through the laboratory analysis of soil and soil pore-water samples. A calibrated model of the system hydrology was developed and used to predict vadose zone leaching, ground-water recharge, and tracer movement. Model predictions were compared with field observations. The water and tracer movement predictions, made possible with intensive monitoring, contribute to the understanding of test system dynamics for both dissipation and PGW studies and provide valuable insight to the field conditions between the discrete sampling events.

Introduction

The draft USEPA guidelines for FIFRA Terrestrial Field Dissipation and Prospective Ground-Water (PGW) studies include soil-moisture monitoring requirements to better understand the study sites hydrology and potential test substance movement (*1, 2*).

The draft Terrestrial Field Dissipation study guidelines state that *"The soil water content can affect the mode of degradation, degree of microbial activity, potential for volatization, plant growth, and potential for movement (up or down in the soil profile). In order to interpret routes and patterns of dissipation of the test substance, the soil-water content needs to be measured on a regular basis to adequately determine the flux of soil water. Various methods of measuring soil water include tensiometers, time domain reflectometry (TDR), neutron probes, gypsum blocks, and direct measurement of the moisture content of the soil samples."* (*1*).

The draft PGW guidelines state that *"soil water content throughout the site should be measured at least monthly"* (*2*). The guidelines specifically suggest that soil-moisture measurements are important for determining the study site water balance, water movement, and solute transport. Measurements should be made near other instrumentation and at least when samples are collected from lysimeters. Instrumentation as listed with the dissipation guidelines are appropriate in PGW studies.

The two study guidelines do not explicitly state that continuous monitoring is essential, although soil-moisture changes are a dynamic process in the agricultural environment and continuous monitoring provides a dramatically different view of the soil environment than could be obtained through discrete sampling methods. The rapid fluctuations in moisture content are evident in the continuous moisture monitoring data presented in the case studies. Two general methods are listed as options for monitoring soil moisture, those that measure potential and those that measure water content. Soil water potential methods include watermark sensors, gypsum block, or tensiometers. Soil water potential measurements require a water content characterization curve to obtain volumetric water content. The characterization curves are soil specific and change in drying and wetting conditions, making direct application of the collected potential data in models or other uses difficult. Time Domain Reflectometry (TDR) and several other technologies offer measurement of soil mosisture content directly. This technology enables continuous, remote monitoring at many depths with reasonable accuracy without calibration.

The case studies presented in this paper present data obtained using TDR technology for soil-moisture measurement. The data presented will make clear that non-continuous monitoring would provide an incomplete picture of an agricultural site receiving rainfall and irrigation water inputs simultaneously with rapid leaching and significant evapotranspiration. The collected data are used to calibrate the hydrology component of the Pesticide Root Zone Model (PRZM, version 3.12) for each site. The calibrated model is used to predict potassium bromide tracer movement which is compared to field observations.

Objective

The objective of this paper is to present the experiences of the authors in implementing the moisture monitoring requirements of the draft Terrestrial Field Dissipation and PGW study guidelines. Instrumentation considerations are discussed including the sensor selection, location, and installation in relation to the two case studies. The case studies illustrate the use of the data in model calibration and the overall understanding of water dynamics at a field study site. This paper is not intended to fully cover all available options for monitoring and modeling or to discuss the theory behind the instrumentation, modeling, or water movement in general.

Case Studies

Methods

Two PGW studies were initiated in the spring of 1999 as part of the registration requirements for a soybean herbicide. Following detailed site characterization, the two sites were instrumented with suction lysimeters, monitoring wells, pipe lysimeters, meteorological monitoring equipment, and a soil-moisture and water-level monitoring system. Each site received applications of the test substance and a conservative potassium bromide tracer as pre-plant bare ground broadcast spray with light incorporation.

PGW Study Site Description

The Southeastern US site is located in North Carolina, on soils identified as Conetoe, sandy loam (loamy, mixed, thermic Arenic Hapludults). The

Midwestern US site is located in Indiana, on soils identified as Tyner, loamy sand (mixed, mesic Typic Udipsamments). Both sites receive irrigation from center pivot systems. Pore-water Br⁻ analysis was performed on samples from suction lysimeters installed at depths 90, 180, 270 cm at both sites with an additional 365 cm depth at the Indiana site. Soil Br⁻ analysis was performed on samples collected from a 0-60 cm depth in 15 cm increments. Characteristics of the two sites follow:

NORTH CAROLINA
Conetoe sandy loam
Arenic Hapludults
- Well drained
- Moderately rapid permeability
- Formed on Atlantic Coastal Plain stream and low marine terraces

Annual Precipitation 130 cm
Depth to water 2.6 to 3.8 m
KBr application 126 kg/ha

INDIANA
Tyner loamy sand
Typic Udipsamments
- Excessively drained
- Rapid permeability
- Formed on Wisconsinan age sandy outwash plains and terraces

Annual Precipitation 102 cm
Depth to water 4.4 to 5.0 m
KBr application 153 kg/ha

Moisture Monitoring Instrumentation

Identical on-site automated weather stations and soil-moisture monitoring systems (Environmental Sensors, Inc., Moisture Point™ TDR, Victoria, BC, Canada) were installed by Waterborne Environmental, Inc. (Leesburg, VA), at each study site along the edge and just outside of the treated area (Figure 1). The weather station/TDR system measures and records soil-moisture as well as meteorological parameters. A Campbell Scientific Inc. (Logan, UT) CR10X data logger is utilized for multiplexer control and data storage. Meteorological sensors are wired directly to the CR10X data logger and measurements are taken every 15 seconds and recorded hourly. Meteorological measurements include precipitation, solar radiation, wind speed and direction, air and soil temperature, and relative humidity. A solar panel, sized appropriately for each location, was installed to ensure a continuously charged battery supply to the data logger and Moisture Point™ system. Data were downloaded remotely on a regular basis via modem and cellular phone.

Soil-moisture measurements were made at three locations along the edge of the treated area (Figure 1). The TDR system measures and records volumetric water content hourly from a series of nine TDR profiling probes. The Moisture Point™ profiling probe system employs conventional TDR technology with a

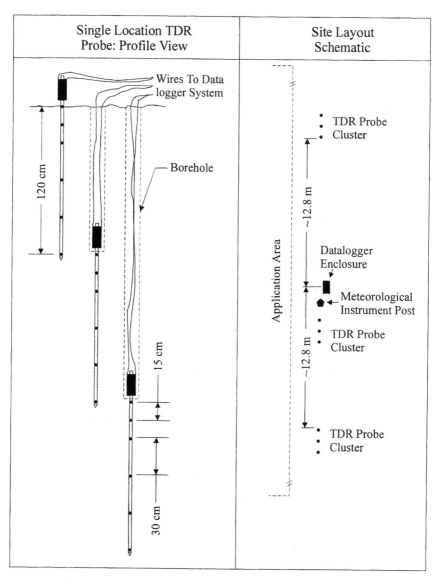

Figure 1. Stacked Profiling Probe Installation and Layout

unique probe design enabling multiple depth readings from a single probe. Diodes, switched on and off by the data logger, are used along the probe to isolate each segment of measurement (3). At each location, three five-segment profiling probes were staggered vertically when installed, thereby providing continuous measurement from the soil surface to a 3.6-meter depth (Figure 1). The profiling probe measures average moisture content (as a percentage of volume) over the depth intervals of 0-15, 15-30, 30-60, 60-90, 90-120 cm when installed vertically. The probes were installed in undisturbed soil either at the soil surface (upper probe) or at the bottom of a borehole (deeper two probes). The small borehole was backfilled with the native materials removed during installation. A solid steel pilot rod was used to create a perfectly sized opening in which the probe is installed. The pilot rod was driven into the ground using a post driver, removed by hand or with a jack. The profile probe was then pushed into the opening left by the pilot rod. A surface installation would be adequate, (0-120 cm depth) for dissipation studies. After sensors were installed, the wires were run on the soil surface in PVC conduit, for rodent and traffic protection, and connected to multiplexers at the central datalogger enclosure (Figure 1).

Measured Data

The electronic instrumentation at the two PGW study sites provided valuable data on site hydrology. The data presented in this paper were summarized to meet the objectives of characterizing water movement at the PGW sites and calibrating a simulation model. The data from each depth interval were averaged, this average was summarized over a 24-hour interval, and the values were multiplied by the representative depth to provide total water in each depth interval. The summarized data for the first 8-months of data collection, are presented for depths of 0-15, 15-30, 30-60, 60-120, and 120-180 cm at the North Carolina site (Figure 2) and at the Indiana site (Figures 3) with two additional depth intervals of 180-240 and 240-360 cm. The deeper depths are not presented for North Carolina because of a rising water table following extreme fall rainfall events. The data at 180 and 360 cm depths showed when and to what depth the soil was saturated, but this period was not appropriate for model calibration and therefore not included. Daily measured rainfall and irrigation (presented as precipitation), as well as calculated evapotranspiration are included graphically (Figures 2 and 3).

The collected data have uses for both qualitative and quantitative analyses. A visual or qualitative analysis of the moisture graphs illustrates to what depth and time interval rainfall and irrigation inputs lead to changes in the moisture profile. For example, at the Indiana site, the measured data at the shallowest depth demonstrates that at every rainfall or irrigation event causes an increase in

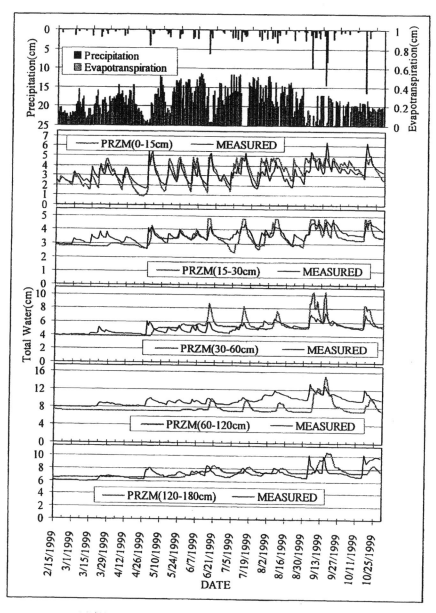

Figure 2. North Carolina Site Precipitation, Evapotranspiration, and Total Soil Water, 0-180 cm (Predicted and Measured)

144

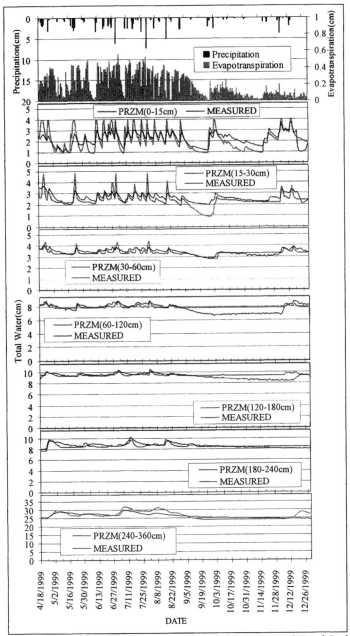

Figure 3. Indiana Site Precipitation, Evapotranspiration, and Total Soil Water, 0-360 cm (Predicted and Measured)

moisture and rapid decrease either through leaching or drying (Figure 3). At the deepest depths, rainfall and irrigation in early May, early June, mid-July, and mid-August leached to the bottom of the measured profile (Figure 3). It is evident from these graphs that a monitoring technique that provides less frequent measurements would not adequately capture the true moisture dynamics at a field study site. The daily summary presented in this paper was selected to correspond to the daily modeling time step. The raw hourly data shows even more rapid changes in the surface horizons than can be seen in the figures.

The final topic of this paper was the quantitative use of the soil-moisture data. The data were used to develop a calibrated model that was used to predict both water and Br⁻ tracer movement. The model predictions were compared to field measurements.

Model Selection

PRZM was selected based on its ability to account for pertinent environmental processes at an appropriate scale and time step for chemical dissipation, and because of preferences for its use by USEPA's Office of Pesticide Programs (4). PRZM-3.12, the current official version as released by the USEPA (5), is a dynamic, compartmental model for use in simulating water and chemical movement in unsaturated soil systems within and below the plant root zone (6). The model simulates time-varying hydrologic behavior on a daily time step, including physical processes of runoff, infiltration, erosion, and evapotranspiration. The chemical transport component of PRZM calculates pesticide uptake by plants, surface runoff, erosion, decay, vertical movement, foliar loss, dispersion and retardation

The PRZM model, in its typical configuration, is limited by the daily time step which requires water to drain to field capacity each day through a "tipping bucket process" to the layer below. Moisture can never exceed field capacity and although recharge predictions may be accurate, they are difficult to verify against measured data because recharge is rarely measured directly and soil moisture predictions are not represented exactly. Utilizing the restricted drainage option with calibration, the "tipping bucket process" limitations are by passed, allowing for verification of predictions to the measured values. Soil, crop, and weather data collected from each site were used to develop the two models which were calibrated using the measured soil-moisture data.

Model Setup and Calibration

The PRZM input files were setup using all available site measured data, with the remainder being estimated from the PRZM3 manual (5) based on the site location, or crop observations. Meteorological data used as inputs include measured rainfall, wind speed, solar radiation, and temperature. Evapotranspiration was estimated using a computer application of the Penman method (7) and on-site measured weather parameters. The soil data include measured texture, organic matter, bulk density, field capacity, and wilting point. The restricted drainage parameter, field capacity and bulk density, were modified within reason to obtain the best fit of total soil moisture at each depth as seen in Figures 2 and 3. The calibrated models were run, adding a potassium bromide application with no degradation, to compare the model predictions of bromide⁻ in soil-pore water and soil samples to field observations (Figures 4 through 7).

Calibration Results and Bromide Predictions

The predicted and measured total water content at multiple depths are displayed in graphical form for the North Carolina (Figure 2) and Indiana (Figure 3) study sites for the initial 8-months of data collection. Evapotranspiration as output from the model and precipitation and irrigation inputs are also displayed. The calibration results visually show a decent fit for the North Carolina site with respect to timing and magnitude at the surface. The predictions are less precise at deeper depths, but still show reasonable magnitude and timing changes at the deepest depths. Many factors including variability at the three measurement locations, limitations of the model, preferential flow, and heterogeneity of soil sample analysis could lead to estimation inaccuracies. The hydrology calibration results for the Indiana site were good with respect to timing at all depths, with magnitude predictions being less accurate in the surface layers for the first half of the simulation.

The calibrated hydrology model was used to predict bromide movement into the soil profile. The predicted and measured soil pore-water concentrations, from eight on-site suction lysimeters, of bromide are presented for the North Carolina (Figure 4) and Indiana (Figure 5) sites as a function of Days After Treatment (DAT). The North Carolina predictions always pass between the observed maximum and minimum concentrations, and generally are close to the average. The Indiana bromide model predictions do not fit as well as for the North Carolina site, but predictions are generally within the range of the observed values, with the exception of the 3-foot values.

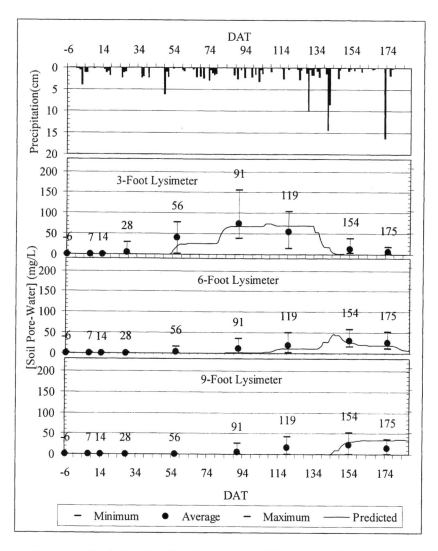

Figure 4. North Carolina Site Predicted and Measured Soil-Pore Water Bromide Concentration

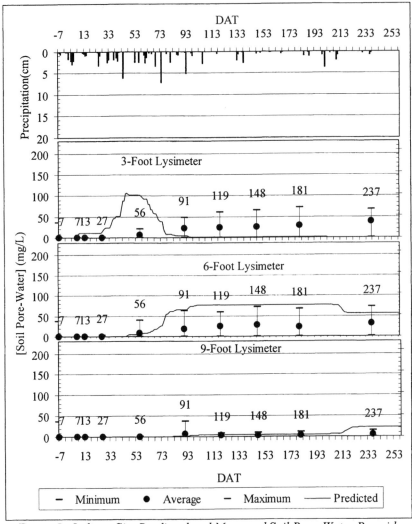

*Figure 5. Indiana Site Predicted and Measured Soil Pore-Water Bromide
Concentration*

The soil concentration of bromide predicted and measured range of values are presented for the North Carolina site (Figure 6) and Indiana site (Figure 7). In North Carolina, the predictions in the 0-15 cm range compared well, with less favorable comparability as the depth increased. The deepest predictions were late with respect to timing, but very close with respect to magnitude. The Indiana site bromide in soil predictions compared well with respect to timing and magnitude at all depths, a slightly different result than was observed in the soil pore-water predictions.

Predicted Water Balance

In addition to the detailed soil-moisture movement and chemical concentrations, model outputs can be used to provide a site water balance. After calibration, when the soil water predictions and observed chemical movement predictions are considered to be reasonable, it can be assumed that the predictions of leaching, evapotranspiration (ET), and runoff losses are also reasonable for the two sites (Figures 8 and 9). The higher runoff as predicted by PRZM at the North Carolina site did not necessary leave the field, rather it may have redistributed within the field and percolated at local depressions.

Summary

Continuous electronic soil-moisture monitoring provides detailed data not obtainable by other methods. Soil-moisture changes can be rapid and discrete monitoring methods would not necessarily capture recharge events. Qualitative examination of continuous moisture data provides information on when and to what depth recharge occurs following every precipitation or irrigation event. Quantitative analysis of moisture data to predict recharge and leaching requires the use of computer simulation models. An accepted USEPA regulatory model, PRZM 3.12 was used to demonstrate the development of a calibrated model to match the continuous TDR soil-moisture measurements of two PGW study sites and used to predict leaching of a conservative tracer and water recharge. The leaching predictions compared reasonably well with measured soil and suction lysimeter data.

Additional components that would add to this paper include the presentation of un-calibrated model results, recommendations for atypical parameter selection (restricted drainage), and a statistical analysis of the goodness of fit of the model predictions both before and after calibration. Potential expansion includes using the calibrated model to make predictions of future events at the PGW site and make predictions of the chemical movement under varying environmental fate

150

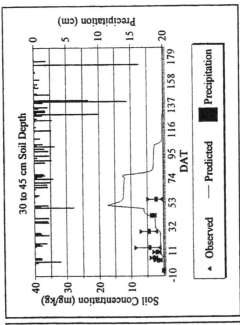

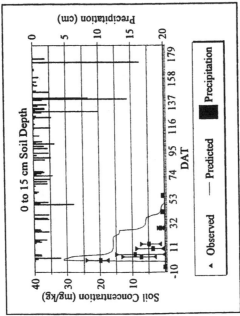

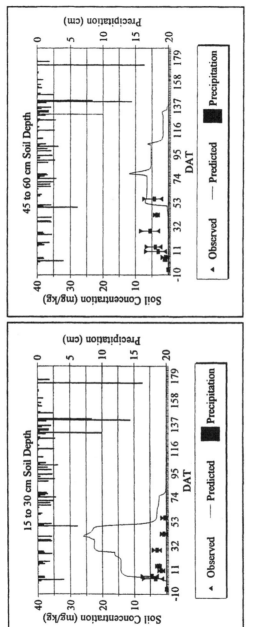

Figure 6. North Carolina Predicted and Measured Soil Bromide Concentrations

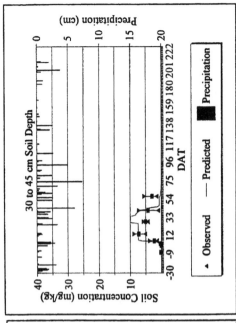

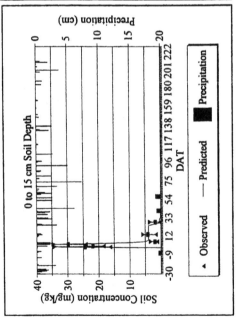

153

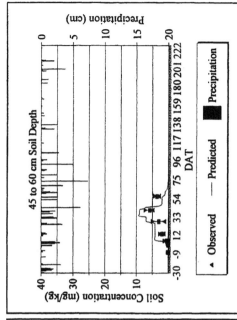

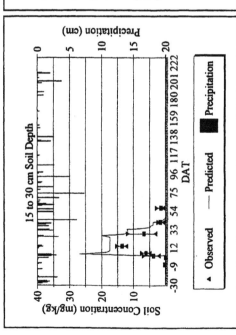

Figure 7. Indiana Predicted and Measured Soil Bromide Concentrations

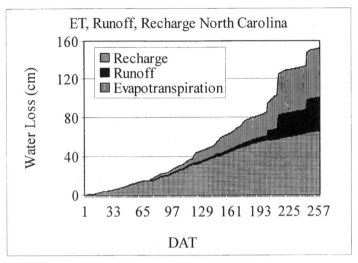

Figure 8. Predicted Water Losses at North Carolina Site

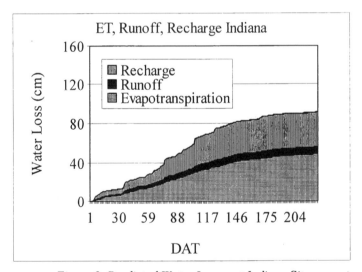

Figure 9. Predicted Water Losses at Indiana Site

and/or weather conditions. For example, the calibrated model could be used to make predictions of extreme events anticipated with annual applications at the site for many years versus the single application used in a PGW study. The objectives of this paper were met in that a case study of the implementation of the moisture monitoring requirements of the draft Field Dissipation and PGW study guidelines were presented highlighting instrumentation, data analysis, and a demonstration of the use of data in modeling to make chemical movement predictions.

References

1. *Draft Proposal for a Test Guidance or Guidance Document, Terrestrial Field Dissipation Studies;* USEPA, Office of Pesticide Programs, Environmental Fate and Effects Division and Health Canada, Pest Management Regulatory Agency, Environmental Assessment Division; September 15, 1998.
2. *Guidance for Prospective Ground-Water Monitoring Studies;* USEPA, Office of Pesticide Programs, Environmental Fate and Effects Division; September 16, 1998.
3. *Moisture Point Model MP-917: Operational Manual;* Environmental Sensors Inc. Victoria, B.C. Canada
4. American Crop Protection Association (ACPA) *Primary, Secondary, and Screening Models for Pesticide Registration*; FIFRA Exposure Modeling Work Group, Washington, DC, April 1995.
5. Carsel, R. F., J. C. Imhoff, P.R. Hummel, J.M. Cheplick, and A. S. Donigian Jr. *PRZM-3, A Model for Predicting Pesticide and Nitrogen Fate in the Crop Root and Unsaturated Soil Zones: Users Manual for Release 3.0*; National Exposure Research Laboratory, Office of Research and Development, U.S. Environmental Protection Agency, Athens, GA (draft) 1998.
6. Mullins, J.A., R.F. Carsel, J.E. Scarbrough, and A.M. Ivery *PRZM-2 A Model for Predicting Pesticide Fate in the Crop Root and Unsaturated Soil Zones: Users Manual for Release 2.0: EPA/600/R-93/046*; Environmental Research Laboratory, Office of Research and Development, U.S. Environmental Protection Agency, Athens, Georgia. 1993.
7. Jones, J.W., L.H. Allen, S.F. Shih, J.S. Rogers, L.C. Hammond, A.G. Smajstrala, and J.D. Martsolf *Estimated and Measured Evapotranspiration for Florida Climate, Crops, and Soils:* Agricultural Experimental Stations, Institute of Food and Agricultural Sciences, University of Florida, Gainesville, Florida, 1984.

Chapter 11

Field Soil Dissipation [^{14}C]ET–751 (Pyraflufen-ethyl) in Bare Ground in California

High Specific Activity Radiolabeled Test Substance and Small Plot Design Afford Excellent Method To Measure Field Dissipation

Fred C. Baker[1], Lilia Estigoy[1], Ella Kimmel[1], Yuji Ikemoto[2], Yukio Kimura[2], and Masao Shigemura[3]

[1]PTRL West, Inc., Alfred Noble Drive, Hercules, Ca 94547
[2]Nihon Nohyaku Company, Ltd., 345 Oyamadacho, Kawachinagano, Osaka 586–0094, Japan
[3]Nihon Nohyaku Company, Ltd., 1-2-5 Nihonbashi, Chuo-Ku, Tokyo 103–8236, Japan

Field soil dissipation studies to support EPA registration of pyraflufen-ethyl (ET-751) were conducted in small outdoor plots (3 x 20 feet) under cotton growing conditions in California using [pyrazole-5-^{14}C] labeled test substance applied at 13.4 g/acre (33.2 g/ha). Use of radiolabeled ET-751 afforded a limit of quantification (LOQ) of < 0.001 parts per million (ppm) for soil total radioactive residues (TRR) determinations and extracted metabolite residues. The $t_{1/2}$ for ET-751 was < 1 day due to rapid hydrolysis to the free acid (E-1). The latter reached its maximum (0.0193 ppm) in 0–3 inch soil segments after 6 hours, but this was short lived ($t_{1/2}$ ~14 days) and residues fell to < 0.010 ppm by 14 DAT. Other identified metabolites were formed from E-1 or ET-751 by reductive cleavage, O-methylation and/or N-demethylation

reactions. Each of the latter metabolites was detected at <<
0.010 ppm in 0–3 inch soil extracts. At least two unidentified
polar metabolites (designated U-1 and U-2) were also
detected. U-2 contained at least two components, which had
chromatographic properties similar to unknown polar
metabolites detected in parallel soil photolysis or aerobic soil
studies. None of the 3–6 inch soil segments contained TRR >
0.010 ppm. TRR in lower soil segments were << 0.010 ppm. A
maximum of 0.0025 ppm E-1 was detected in 3–6 inch soil
extracts (14 DAT) and U-2 reached 0.0038 ppm in 28 DAT 3–
6 inch soil. By 359 DAT all metabolite residues were 0.001
ppm, or less, in 0–3 inch and 3–6 inch soil extracts. The
degradation of parent and formation of known metabolites in
the dissipation study and a parallel aerobic soil metabolism
study was similar both qualitatively and quantitatively.

ET-751   E-1

Introduction

Pyraflufen-ethyl, ethyl 2-chloro-5-(4-chloro-5-difluoromethoxy-1-
methylpyrazol-3-yl)-4-fluoro-phenoxyacetate, is a broad spectrum post-
emergence contact herbicide developed by Nihon Nohyaku Co. Ltd. Its mode of
action is by inhibition of protoporphyrinogen-IX oxidase. Uses include selective
control of broadleaf weeds such as *Galium aparine* (cleavers), *Matricaria
inodora* (scentless false chamomile), *Lamium purpureum* (red dead nettle) and
Stellaria media (chickweed) in cereals (*1*). Use as a defoliant of cotton (2.5–5.0
g/ha) and potatoes (15–30 g/ha) is also intended.

The dissipation of ^{14}C labeled test substance, and its degradates, in a soil
representative of a cotton growing region, was conducted to partially satisfy the
requirements of FIFRA Subdivision N Guideline 164-1 (Field Dissipation for

Terrestrial Use). Because of the low treatment rate (nominally ~ 30 g a.i./hectare, ~ 12 g a.i./acre) the study utilized [pyrazole-5-^{14}C]ET-751 of high specific activity (324,120 dpm/µg; 60.2 mCi/mmol). This was to facilitate detection of anticipated low level residues as well as to identify, quantify and monitor formation and decline of ET-751 metabolites for the study duration. The use of high ^{14}C specific activity test substance to conduct a small plot field dissipation study facilitates accounting of mass balance for non-volatile residues. Furthermore, determination of total radioactive residues (TRR) in soil is an excellent screening procedure to determine which soil segments require additional analysis for metabolite i.d. and quantification.

Previously, the dissipation and mobility of ET-751 and its metabolites E-1, E-2 and E-3 were studied in four soils under field conditions in France, Germany and the United Kingdom (2, 3). In those studies, unlabeled ET-751 was used at a rate of 200 g a.i./ha and analysis was by LC/MS/MS. The limit of detection was ~ 0.01 ppm. In the former study (2; spring application) the 50% Decline Time (DT_{50}) for ET-751 was estimated to be about 1–7 days in 0–5 cm soil segments. E-1 concentrations ranged from 0.05–0.27 ppm in 0–5 cm soil initially and reached a maximum of 0.48 ppm 14 DAT. Levels of E-1 declined steadily thereafter to < 0.01–0.04 ppm after 12 months. Some leaching of E-1 into 5–10 cm soil segments occurred but only on one occasion into 10–15 cm horizons. The $t_{1/2}$ for E-1 was estimated to be 11–44 days. Metabolite E-2 was detected on only two occasions (3 months after application) at 0.01 ppm, a level close to the limit of detection of the analytical method. E-3 was not detected until 7 days–2 months following application. After this time, residues of 0.01–0.05 ppm were regularly detected in 0–5 cm soil up to one year after treatment. E-3 was also found in 5–10 cm horizons at 0.01 ppm and on one occasion in the 10–15 cm horizon. Similar results were obtained from the autumn application study (3).

Experimental

Purity of the test substance was established by HPLC analysis (pre-application) and analysis of a retain sample of the formulated dosing solution (post-application). Results of the purity checks were 96.2% (pre-application) and 95.9% (post-application).

The field phase of this study was carried out by Excel Research Services, Inc., 3021 West Dakota Road, Suite 110, Fresno, CA 93722. Study plots (containing Atwater loamy sand; pH 7.2 in 0-6" segment) were located at the Excel Madera Research Station, CA 93637. Two treated (A, B) and one control plot were used. Each plot measured 3 x 20 feet and was subdivided into 60 subplots (1 x 1 foot). Plots were bordered by 2 x 6 inch wooden planks which were marked at 1 foot increments to delineate the positions of the subplots for coring purposes. The control plot was at a distance of > 200 feet from the radioactive plots. The ^{14}C-plots were located within a designated area surrounded by a chain link security fence, equipped with locked gates and marked with radioactive caution signs. Plots were marked using a corner stake with study number, treatment (control, treated A and treated B), date and initials. The untreated control plot was not fenced.

Five paper target pads (2 inches x 2 inches) were placed in each plot at selected intersects. Radiolabeled pyraflufen-ethyl was applied to bare soil as a 2.5% EC formulation using a CO_2 backpack sprayer and a single pass. The target rate for application was 30 g/ha. The actual application rates were 32.8 g/ha and 33.5 g/ha (average of 33.2 g/ha, 13.4 g/acre). Soil cores were removed with a Giddings™ hydraulic sampler mounted on a trailer. Soil core harvest details are shown in Table I.

Table I. Soil Core Harvest Details

Date	Days after Treatment (DAT)	Plot Sampled	Cores Taken[1]
5/26/98	-1	Control, A, B	0–36 inch (3x control, 1xA, 1xB)
5/27/98	0	Control, A, B	0–6 inch (1x control, 5xA, 5xB)
5/27/98	0.25	Control, A, B	0–6 inch (1x control, 3xA, 3xB)
5/28/98	1	A, B	0–6 inch (3xA, 3xB)
5/30/98	3	A, B	0–36 inch (3xA, 3xB)
6/3/98	7	A, B	0–36 inch (3xA, 3xB)
6/10/98	14	Control, A, B	0–36 inch (1x control, 3xA, 3xB)
6/24/98	28	A, B	0–36 inch (3xA, 3xB)
7/27/98	61	A, B	0–36 inch (3xA, 3xB)
8/25/98	90	A, B	0–36 inch (3xA, 3xB)
9/25/98	121	Control, A, B	0–36 inch (1x control, 3xA, 3xB)
11/24/98	181	A, B	0–36 inch (3xA, 3xB)
1/22/99	240	A, B	0–36 inch (3xA, 3xB)
3/23/99	300	A, B	0–36 inch (3xA, 3xB)
5/21/99	359	Control, A, B	0–36 inch (1x control, 3xA, 3xB)

[1] 0–6 inch = 2.25 inch diameter; 6–36 inch = 1.75 inch diameter.

Plots were irrigated as necessary to maintain 120% of normal monthly precipitation based on the 10 year historical average, or at least 120% of typical crop water needs for the general area, whichever was greater. The amount of water a cotton crop in California would typically need was based on information provided by the University of California Cooperative Extension (Tulare County).

Plots were watered by sprinklers. The total amount of water (rainfall and irrigation) received by the plots was 47.51 inches, which was 104.8% of the targeted rate and 125.7% of typical crop water needs.

Target pads and soil cores were shipped frozen to PTRL West for analysis. Target pads were extracted with acetonitrile (ACN) for LSC analysis. Cores were cut into 6 inch segments, and 0–3 inch and 3–6 inch segments were cut from 0–6 inch segments. Soil segments were homogenized in a food processor and aliquots were removed for moisture determination and for radioassay by combustion (3 replicates). Total radioactive residues (in ppm) were calculated based on dry soil weight for each core. At each harvest interval (see Table 1) corresponding core segments containing TRR $\geq$ 0.003 ppm were combined (50 gram subsamples of each homogenized core) to form new composite samples for plot A and plot B, respectively. The composite samples were homogenized and aliquots removed for wet/dry weight determination and for further TRR determination.

Composite soil samples were extracted as shown in Scheme 1 (method modified from Reference 4). LSC of aliquots of soil extracts allowed calculation of ^{14}C dpm/g dry soil. Soil extracts from composite soil samples were combined with reference standards and analyzed by gradient HPLC (0.01% formic acid in water/acetonitrile), with ^{14}C and UV detection. The percentage ^{14}C corresponding to each metabolite allowed calculation of ^{14}C dpm associated with each metabolite/g dry soil. The specific activity of metabolites was calculated from the specific activity of ET-751 (324,120 dpm/μg) and the appropriate molecular weight conversion factor. Thus, residue levels of each metabolite could be calculated on a μg/g (ppm) dry soil.

Calculations

Soil TRR Limits of Quantification (LOQ) =
$$\frac{2 \times \text{bkgd}\,(40\,\text{dpm})}{\text{sp. act. ET-751}} = 0.123 \text{ ng ET-751}$$

If 1 g soil combusted then LOQ = $\dfrac{0.123\,\text{ng}}{\text{g}}$ = ~ 0.0001 ppm

Metabolite ppm =
$$\frac{^{14}C\,\text{dpm extracted}}{\text{g dry soil}} \times \frac{\%\,\text{Metabolite (HPLC)}}{100} \times \frac{1}{\text{Metabolite sp. act. (dpm/}\mu\text{g)}}$$

Metabolite specific activity (dpm/μg) =
$$\text{ET-751 sp.act. (324,120 dpm/}\mu\text{g)} \times \frac{\text{Mol. Wt. ET-751}}{\text{Mol. Wt. Metabolite}}$$

Scheme 1

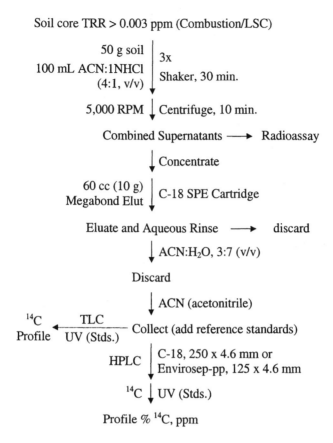

Soil core TRR > 0.003 ppm (Combustion/LSC)

50 g soil
100 mL ACN:1NHCl   3x
(4:1, v/v)   Shaker, 30 min.

5,000 RPM   Centrifuge, 10 min.

Combined Supernatants ⟶ Radioassay

Concentrate

60 cc (10 g)
Megabond Elut   C-18 SPE Cartridge

Eluate and Aqueous Rinse ⟶ discard

ACN:H₂O, 3:7 (v/v)

Discard

ACN (acetonitrile)

¹⁴C ⟵ TLC
Profile   UV (Stds.) ⎯ Collect (add reference standards)

HPLC   C-18, 250 x 4.6 mm or
Envirosep-pp, 125 x 4.6 mm

¹⁴C   UV (Stds.)

Profile % ¹⁴C, ppm

Results

Target Pad Analysis/Validation

The mean level of ^{14}C extracted from target pads was 2.2394 x 10^6 dpm/pad from plot A, and 2.2112 x 10^6 dpm/pad from plot B. Based on target pad area of 4 square inches and the dimensions of the plot and application rate, the level of extracted radiolabel was 81.3% and 78.4% of theoretical for plots A and B, respectively. The recovery of ^{14}C ET-751 from target pads was acceptable considering that test substance was applied by spraying the plot and some spray drift could be anticipated. Recovery of radiolabel from target pads spiked in the laboratory with ^{14}C ET-751 was 102%.

Soil Total Radioactive Residues (TRR)

Figure 1 shows the TRR decline curves determined on a dry soil basis for 0–3 inch and 3–6 inch soil segments (horizons) throughout the study duration.

Mean TRR in 0–3 inch soil segments ranged from 0.025 ppm post-application to 0.027 ppm 6 hours after application. Subsequent levels determined were 0.023 ppm, 0.023 ppm and 0.027 ppm at 1 DAT, 3 DAT and 7 DAT, respectively. These data are in close agreement with calculated levels based on application rate. After 7 DAT residues declined and by 61 DAT mean level was < 0.01 ppm (0.008 ppm). By 359 DAT TRR in 0–3 inch soil was 0.005 ppm. Mean TRR in 3–6 inch soil segments was 0 ppm post-application and reached a maximum of 0.008 ppm by 14 DAT. Thereafter, residues declined to 0.003 ppm (by 300 DAT and 359 DAT). Residues in 6–12 inch and 12–18 inch soil segments were generally undetectable, or < 0.002 ppm. The maximum TRR observed in 6–12 inch soil was 0.002 ppm (28 DAT), and there were no detectable residues in 12–18 inch soil. The $t_{1/2}$ for TRR in 0–3 inch soil was ~ 28 days. TRR data determined for composite plot A and B soil samples were very similar to those determined for individual cores.

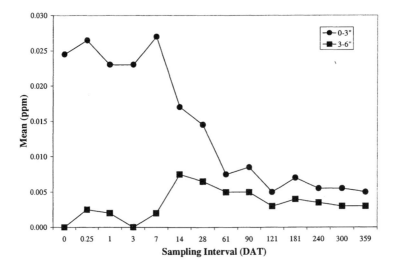

Figure 1. Decay Curve for Mean TRR in 0–3 and 3–6 Inch Soil Segments

Formation and/or Decline of Pyraflufen-ethyl and Metabolites (0–3 Inch and 3–6 Inch Soil) (see Figure 5 for chemical structures)

Pyraflufen-ethyl (ET-751) accounted for only 17.41% (0.0035 ppm) and 15.28% (0.0039 ppm) of the radiolabel in extracts from plot A and B post-application soil, respectively. The hydrolysis product (E-1) accounted for most of the balance of the radiolabel: 76.09% (0.0143 ppm) in plot A; 79.24% (0.0189 ppm) in plot B. A low level of E-2 was also detected in post-application soil extracts. Extracts from 6 hours and 1 DAT soil showed similar ^{14}C profiles except for the minor components. Figure 2 shows the radiochromatogram and UV chromatogram from analysis of plot A 1 DAT 0–3 inch soil extract. E-1 accounted for 82.5% (0.0141 ppm) and ET-751 accounted for only 6.6% (0.0012 ppm). Low levels of polar metabolites designated U-1 (1.80%, 0.0003 ppm) and U-2 (4.00%, 0.0007 ppm) were also detected, as well as 0.0002 ppm E-2 and E-9. By 3 DAT ET-751 was undetectable in plot A soil and accounted for only 2.30% (0.0006 ppm) in plot B soil extract. E-1 began to decline 7 DAT while U-1, U-2 and E-9 increased. Figure 3 shows the metabolite radiochromatogram profile in 28 DAT plot B 0–3 inch soil extracts. E-3 had become the predominant metabolite with concommitant decline of E-1 and appearance of E-9, E-2 and E-11, as well as polar unknown metabolites. Figure 4 shows the decline curve for E-1 in 0–3 inch soil extracts.

Tables II and III summarize the results from HPLC analysis of all soil extracts and shows the calculated residue levels (ppm) of metabolites for the study duration. Assignments of ET-751 and major metabolites based on RPLC

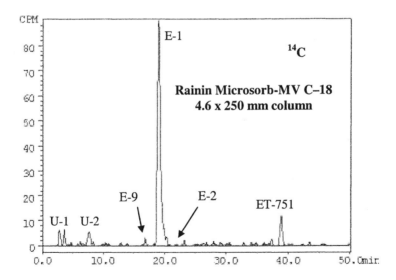

Figure 2a. HPLC Radiochromatogram from Analysis of 1 DAT A 0–3 Inch Soil

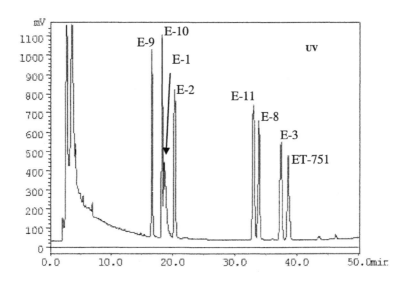

Figure 2b. HPLC UV Chromatogram from Analysis of 1 DAT A 0–3 Inch Soil

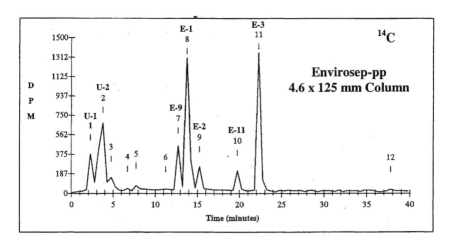

Figure 3. HPLC Radiochromatogram from Analysis of 28 Plot B 0–3 Inch Soil

Table II. Summary of Metabolite ppm in Extracts from 0–3 Inch Soil.

Time	Metabolite							
(DAT)	ET-751	E-1	E-2	E-3	E-9	E-11	U-1	U-2
0	0.0037	0.0166	0.0003	0.0000	0.0000	0.0000	0.0000	0.0000
0.25	0.0041	0.0198	0.0004	0.0000	0.0001	0.0000	0.0001	0.0003
1	0.0022	0.0167	0.0001	0.0000	0.0001	0.0000	0.0004	0.0007
3	0.0003	0.0170	0.0000	0.0000	0.0006	0.0000	0.0004	0.0021
7	0.0000	0.0169	0.0000	0.0000	0.0011	0.0000	0.0003	0.0053
14	0.0002	0.0073	0.0007	0.0009	0.0013	0.0001	0.0010	0.0020
28	0.0000	0.0030	0.0005	0.0027	0.0010	0.0005	0.0010	0.0024
61	0.0000	0.0006	0.0002	0.0014	0.0004	0.0005	0.0008	0.0008
90	0.0000	0.0005	0.0001	0.0027	0.0002	0.0010	0.0007	0.0003
121	0.0000	0.0003	0.0001	0.0012	0.0001	0.0007	0.0005	0.0004
181	0.0000	0.0003	0.0002	0.0016	0.0001	0.0008	0.0004	0.0005
240	0.0000	0.0004	0.0001	0.0013	0.0000	0.0008	0.0005	0.0002
359	0.0000	0.0002	0.0001	0.0010	0.0001	0.0007	0.0003	0.0005

Table III. Summary of Metabolite ppm in Extracts from 3–6 Inch Soil.

Time	Metabolite							
(DAT)	ET-751	E-1	E-2	E-3	E-9	E-11	U-1	U-2
14	0.0000	0.0025	0.0005	0.0003	0.0002	0.0000	0.0009	0.0028
28	0.0000	0.0005	0.0001	0.0005	0.0002	0.0001	0.0002	0.0038
61	0.0000	0.0003	0.0002	0.0005	0.0001	0.0003	0.0002	0.0020
90	0.0000	0.0001	0.0001	0.0015	0.0001	0.0006	0.0006	0.0005
121	0.0000	0.0001	0.0000	0.0007	0.0000	0.0004	0.0004	0.0003
181	0.0000	0.0001	0.0000	0.0009	0.0000	0.0006	0.0003	0.0006
240	0.0000	0.0001	0.0001	0.0012	0.0000	0.0003	0.0005	0.0003
359	0.0000	0.0001	0.0001	0.0007	0.0001	0.0005	0.0003	0.0004

Metabolite ppm corrected for molecular weight and appropriate specific activity

For U-1 and U-2 the specific activity of ET-751 was used: 324,120 dpm/microgram.

co-elution with reference standards was further confirmed by 2-D TLC (silica gel). TLC analysis of U-1 and U-2 gave evidence of at least two polar metabolites with some breakdown to a more polar fraction which did not migrate on TLC. Furthermore, 2-D TLC analysis of a U-2 isolate produced one major radiolabeled zone plus a more polar region tentatively designated as U-1. It is likely that both U-1 and U-2 contain more than a single metabolite.

Isolated U-1 and U-2 fractions from 0–3 inch soil 28 DAT plot A were co-injected with samples of soil extract from aerobic soil study (*4*). U-1 contained at least one radiolabeled component that was more polar than radiolabeled metabolites designated as U-1 and U-2 in the aerobic soil study. U-2 appeared to contain radiolabeled components associated with aerobic soil Unk 1 ($R_T \sim 10$ minute) and a proposed photolysis unknown ($R_T \sim 15$ minute) from a parallel soil photolysis study (*5*).

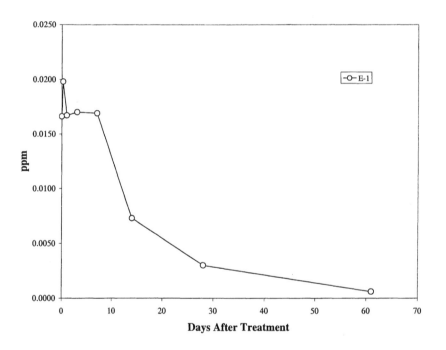

Figure 4. Formation/Decline Curve for E-1

Leaching of Metabolites

Very little leaching occurs from 0–3 inch to 3–6 inch soil segments. Maximum TRR in 3–6 inch soil segments was 0.009 ppm and occurred 14 DAT (plot A). The average maximum metabolite residue observed in 3–6 inch soil segments was 0.0038 ppm U-2 (28 DAT, see Table III). TRR determination in 6–12 inch and 12–18 inch soil segments showed that total residues were $\ll 0.01$ ppm. In 6–12 inch soil the maximum residue was 0.002 ppm in plot A soil 28 DAT. All other 6–12 inch soil segments contained 0.001 ppm or no detectable residues, and 12–18 inch soil segments contained no detectable residues. Thus, essentially no leaching of metabolites occurs below the 0–6 inch layer.

TRR was determined to a depth of 36 inches at 121 DAT and 359 DAT. All 6 inch cores between 18 inches and 36 inches contained essentially undetectable residues.

Pyraflufen-ethyl and E-1 Half-Life and Metabolism

The half-life of ET-751 was exceedingly short and was estimated as < 1 day; calculations including theoretical t_0 ET-751 ppm would produce values < 1 hour. In the post application samples (1–2 hours from application to coring), ET-751 had undergone substantial hydrolysis to its acid metabolite E-1. The DT_{90} was < 1 day. The half life of E-1 was 14 days and the DT_{90} was about 61 days. Figure 5 shows a proposed metabolism scheme for ET-751 in soil.

Storage Stability

Storage stability was established by re-extracting composite soil samples of 14 DAT and 90 DAT 0–3 inch Plot A and B after samples had been stored in the freezer for 289 and 225 days, respectively, since the initial extraction date (this corresponded to a total of 314 days and 240 days storage, respectively, at PTRL West, Inc.). The extracts were subjected to HPLC analysis. No significant difference in metabolite distribution was observed from the first extraction on July 13, 1998 (14 DAT) and September 15, 1998 (90 DAT) to re-extraction on April 28, 1999 (14, 90 DAT).

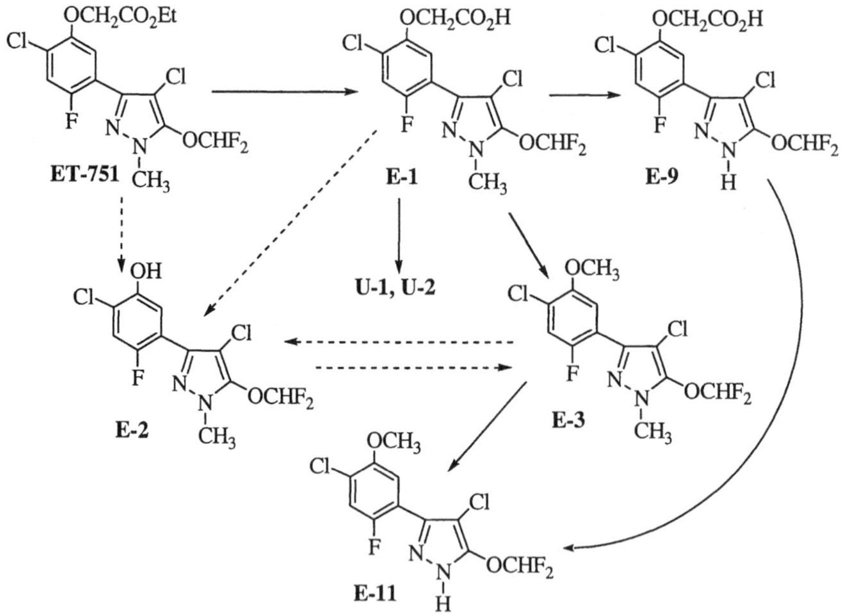

Figure 5. Metabolism Scheme for ET-751 in Soil

Note: Unlabeled reference standard E-8 (ethyl 2-chloro-5-(4-chloro-5-difluoromethoxy-1H-pyrazol-3-yl)-4-fluorophenoxyacetate) was added to soil extracts prior to HPLC along with other reference standards but radiolabeled E-8 was not detected.

Conclusions

- The half-life of pyraflufen-ethyl (ET-751) was < 1 day in a slightly basic (pH 7.2) loamy sand soil in California. The DT_{90} was < 1 day.
- The estimated Decline Time 50 (DT_{50}) for E-1 was 14 days, and the DT_{90} was about 61 days. E-1 was the only metabolite > 0.01 ppm.
- Total radioactive residues (TRR) in 0–3 inch soil decreased to < 0.01 ppm by 61 DAT.
- E-3 was a metabolite of E-1. Residues of E-3 reached a maximum of 0.0032 ppm in 90 DAT 0–3 inch soil.

- U-1 and U-2 were polar unidentified metabolites. U-1 reached only 0.0010 ppm 14 DAT and 28 DAT. U-2 reached 0.0053 ppm in 0–3inch 7 DAT soil then declined rapidly.
- By 359 DAT the TRR had declined to 0.005 ppm in 0–3 inch soil, 0.003 ppm in 3–6 inch soil and 0.001 ppm in 6–12 inch soil.
- Very little leaching occurred; maximum TRR in 6–12 inch segment = 0.002 ppm.
- This small plot/^{14}C label study afforded a highly sensitive and simple method to quantify dissipation of ET-751 and to evaluate the degradation/leaching pattern under natural conditions.

References

1. The Pesticide Manual, 10th Edition, edited by C. Tomlin, Crop Protection Publications, 1004.
2. Benwell, L.; Burden, A. N. ET-751 SC (Containing 20 g ET-751/L): Dissipation from Four Field Soils Following Spring Application. Report 608/26-1016a, Covance Labs, (Formerly Corning Hazelton) UK.
3. Burden, A. N.; Dunn, L. ET-751 SC (Containing 20 g ET-751/L): Dissipation from Four Field Soils Following Autumn Application. Report 608/28-1016, Covance Labs (formerly Corning Hazelton) UK.
4. Shepler, K. Aerobic Soil Metabolism of [^{14}C]ET-751, PTRL West, Inc. Study 745W, September 27, 2000.
5. Concha, M. Photodegradation of [^{14}C]-ET-751 in/on Soil by Artificial Light, PTRL West, Inc. Study 783W, September 19, 2000.

Chapter 12

Diazinon in Apple Orchards: Dissipation from Vegetation and Exposure to Non-Target Organisms

George P. Cobb[1], Eric H. H. Hol[1], Larry W. Brewer[2], and Catherine M. Bens[1]

[1]The Institute of Environmental and Human Health and the Department of Environmental Toxicology, Texas Tech University, P.O. Box 41163, Lubbock, TX 79409–1163
[2]EBA, Inc., P.O. Box 2005, Sisters, OR 97759–3210

Comprehensive residue determinations were made for leaves, grass, and other biota from apple orchards following repeated applications of diazinon (D-z-n® 50W). Orchards in Washington and Pennsylvania served as study areas representing different climatic zones and agricultural practices. Mean diazinon application was measured to be 3.0 to 3.1 kg AI/ha in over 230 samples from orchards in the states studied. Vegetation was collected in four orchards from each state on days 0, 4, and 12 post application. Diazinon dissipation from vegetation was measured following repeated applications in Washington and Pennsylvania orchards. Residue dissipation was described by a simple exponential function r >0.93). The individual degradation rates measured in each field provide a distribution of diazinon presence that can be used for probablistic exposure assessments. It is also worthwhile to note that no significant exposure to earthworms or avian species occurred more than 4 days post application. Significant exposure to these non-target species occurred when at least 19% of applied residue remained on vegetation.

INTRODUCTION

Diazinon (O,O,-dimethyl-O-2-isopropyl-4-methyl-6-pyrimidyl-phosphorothioate) is a widely used insecticide in agricultural and residential settings (1). When used to protect orchard crops, diazinon is formulated as the wetable powder D-z-n® 50W (EPA Registry No. 100-460). Diazinon has a low mammalian toxicity, with LD50s ranging from 300 mg/kg for rats to over 1000 mg/kg for sheep. However, diazinon is more toxic to birds than to mammals (2). Avian LD50s range from 2.5 mg/kg for turkeys *(Meleagris gallopavo)*(3) to 213 mg/kg for European starlings *(Sturnus vulgarus)*(4). Adverse effects have been noted following diazinon ingestion by free ranging waterfowl and passerines (5-10). The wide range of LD50s and documented effects on wildlife emphasize the need for accurate measurements of diazinon presence in ecosystems occupied by birds.

The data presented in this dissipation study are part of a larger quantification of diazinon exposure and effect in avian species. Chemical analyses have provided good measures of pesticide exposure to avian populations (5,9,11-16). These analyses are particularly powerful when used in comprehensive ecotoxicology evaluations (17,18).

METHODS

Site Selection: Study areas were located in south central Pennsylvania (PA), near Biglerville in Adams County, and in Central Washington (WA), near Brewster in the Methow River Valley. Several criteria were used to select orchards for this study (19). Habitat diversity, isolation from other orchards, and orchardist cooperation were primary considerations when studying chemical behavior in a high value crop such as apples. Four treatment and one control site were randomly selected in each state and sampled for residues. Treatment orchards selected in PA included PA-02, PA-05, PA-19, and PA-21. Treatment orchards selected in WA included WA-03, WA-10, WA-14, and WA-15. Due to monetary considerations, control sites were not all orchards thus matched vegetation sampling did not occur.

Diazinon Application: Pesticide was applied by an air blast sprayer pulled by a tractor. Application began in March and continued until early- to mid-July. The first one or two sprays each season typically contained diazinon in an oil mixture (19). Thereafter, aqueous emulsions are used. Output of the sprayers and equipment ground speed were calibrated to provide a nominal application rate of 3.4 kg AI/ha.

The first application (A1) in each orchard was a dormant spray containing diazinon in a superior oil. A2 in PA was also dormant, but weather conditions precluded a second dormant spray in WA orchards. During the time interval separating dormant and foliar applications, honeybees were introduced to the orchards for pollination and no spraying occurred. Foliar applications resumed approximately three weeks after pollenation. Four foliar sprays were performed at 2-wk. intervals (19). During these applications 233 samples were taken across the four treatment fields to estimate application rate in PA, and 244 samples were collected in WA. Samples were collected from tanks of application equipment before, during, and after application in each field on each day. These data were coupled with measured ground speed of application equipment to estimate application rate.

Environmental Sampling: Vegetation samples were collected from 12 permanent collection stations at each of four randomly selected orchards in each state. The twelve permanent collection stations were located randomly within each orchard. Environmental samples were collected from each station before the first diazinon application to serve as controls. We then sampled three times following each diazinon application: within 2 hours of application (D0), 4 days post-application (D4), and 12 days after application (D12). Sampling included: apple leaves (LV), under story vegetation collected beneath the drip ring (DR) of the canopy, and under story vegetation from the approximate center of the treerow (TR). Each vegetation sample was uniquely numbered and stored in plastic Ziploc® bags (Cobb et al., 2000). Samples were not composited across collection stations, providing twelve samples from each orchard at each time point.

Each LV sample contained approximately 25 g of material which was collected from 8 different branches of a single tree. Leaves matured earlier in WA and were available beginning with application A1. However, leaves developed more slowly in PA. Therefore, no leaves were collected following application A1 from PA orchards, and few were collected following A2. Each DR or TR sample contained all vegetation collected within a 15 cm x 15 cm quadrant. All vegetation above the soil surface was collected, including dead vegetation. Standing vegetation was removed with shears, and fallen vegetation was removed with small rakes.

Earthworms were collected in 1 m^2 quadrants (1m x 1m) by digging to a depth of approximately 8 inches and if this failed to produce earthworms, soil flooding was attempted. All earthworms found in a given quadrant were analyzed. Soil particles adhering to the exterior surfaces of earthworms were not removed, and GI tract depuration was not performed. All diazinon data presented for avian species represent residues in GI tracts from carcasses found in the field (19). Samples were frozen until shipped to the laboratory facilities. Samples were shipped in coolers with dry ice and were returned to freezers immediately upon receipt at the analytical laboratories.

Chemical Analysis: Samples were analyzed for diazinon residues using solvent extraction followed by gas chromatography-flame photometric detection (GC-FPD). Analytical procedures use internal spikes for recovery and periodic blank analyses. Tissues were extracted with 1:1 hexane:acetone mixture. Extracts were then concentrated with Kuderna-Danish (KD) evaporators and eluted through 10 g Florisil. Each sample was loaded on a separate column and 25 ml of hexane was added. The hexane was followed by elution with 25 ml of 25% ether in hexane. The latter fraction was collected and concentrated by final KD to 2 ml. Concentrates were stored, frozen until analyses were performed. Analytical separations were performed with a Hewlett-Packard 5890 gas chromatograph equipped with a 30 m x 0.53 mm DB-210 column and helium carrier gas. The temperature program began at 140 C for one minute and ramped at 5 C to 200 C. The FPD was operated in the phosphorous mode. Limits of detection for vegetation and animal tissues were 0.2 µg/g and 0.007 µg/g, respectively. Recoveries from spiked samples were 82% (19).

RESULTS

Diazinon Applications: Diazinon 50W applications occurred from 5 April to 6 July 1989 in test orchards in PA. Applications in WA occurred from 8 April to 7 July 1989 (Table 1). Although measured application rates varied among orchards, the application rates on individual orchards were precise. The mean application rate for all diazinon applications to test orchards in each study region was within 3% of the nominal rate of 3.4 kg AI/ha. Residue analyses of tank samples from confirmed consistent application rates of 3.16±0.20 (mean±SE) and 3.06±0.14 kgAI/ha in PA and WA, respectively. The average application rate on a given field on a given day show many values near the mean application rate with a few concentrations that are substantially removed from the mean (Figures 1&2).

    Vegetation Samples: LVs were first available in PA on A2D12 (Application 2 Day 12 post application). LVs were available on trees in WA by A1D4. Diazinon residues were found on leaves at predictable concentrations throughout the study period as indicated by coefficients of variation ranging from 3% to 6% for data collected on given dates post-application (Table 2). Mean diazinon concentrations in DR and TR samples were compared on D0, D4, and D12. Concentrations differences in DR and TR samples on a given day were 24%, 11% and 4%, respectively. These data indicate that over time diazinon concentrations in DR converge with concentrations in TR samples. In PA, diazinon residues from DR and TR were within 67% of one another on 18 of the sampling days. The highest diazinon concentrations in all orchards were recovered from apple leaves, which is consistent with the spray patterns, which target the tree canopy.

    In PA, residues from leaves on D0 were more than twice the concentration found on under story vegetation (Table 2). By D4 in PA, respective diazinon

Table 1. Julian dates of diazinon application to apple orchards in Washington and Pennsylvania.

Application	WA Dates	PA Dates
1	99-104	93-97
2	128-131	107-110
3	142-145	142-145
4	157-163	152-159
5	172-178	167-173
6		179-187

Table 2.　　　Diazinon concentrations (mean±SE) on vegetative matter from apple orchards in Washington and Pennsylvania.

			Diazinon Concentrations (μg/g)		
State	DPA[a]	N[b]	Tree Row	Drip Ring	Apple Leaves
WA	0	204	62.3 $\pm$ 1.6	47.9 $\pm$ 1.1	306 $\pm$ 10
WA	4	240	11.9 $\pm$ 0.5	10.6 $\pm$ 0.6	79.1 $\pm$ 3.3
WA	12	228	2.4 $\pm$ 0.1	2.3 $\pm$ 0.1	27.6 $\pm$ 1.0
PA	0	180	66.4 $\pm$ 4.0	65.9 $\pm$ 3.7	149 $\pm$ 5.0
PA	4	180	6.9 $\pm$ 0.7	12.5 $\pm$ 1.0	38.0 $\pm$ 1.1
PA	12	240	2.1 $\pm$ 0.3	2.8 $\pm$ 0.2	14.5 $\pm$ 0.9

[a] - Day Post Application

[b] - N= Total number of each vegetation type analyzed across 4 orchards within a given state.

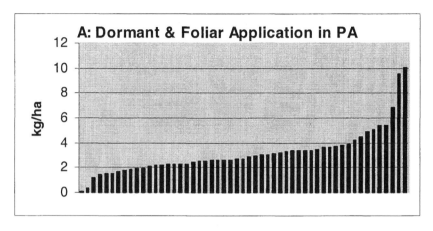

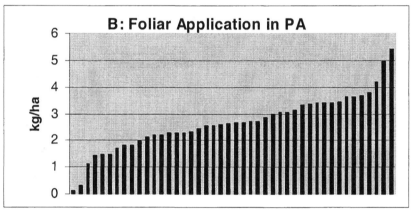

Figure 1. Distributions of measured diazinon application rates in Pennsylvania apple orchards: A) dormant and foliar applications; B) foliar applications only. Multiple bars are present for applications to some orchards where diazinon tanks were refilled during application. The mean application rate is 3.14 kg/ha. Note: many of the outliers are due to dormant spraying.

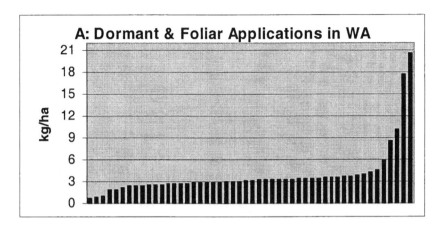

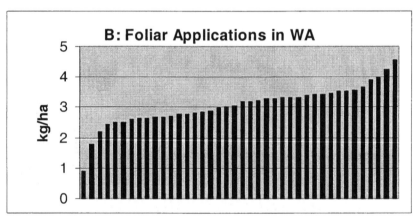

Figure 2. Distributions of measured diazinon application rates in Washington apple orchards. A) dormant and foliar applications; B) foliar applications only. Multiple bars are present for applications to some orchards where diazinon tanks were refilled during application. The mean application rate is 3.04 kg/ha. Note: many of the outliers are due to dormant spraying.

residues from DR and TR were 19% and 10% of concentrations for D0, and those from LVs were 26% of initial concentrations. Residues in the under story vegetation were nearly equal except on D4 when mean concentrations from DRs were nearly twice those obtained from TRs. By D12, diazinon residues from DRs and TRs were reduced to approximately 4% of initial concentrations, and those from LVs were approximately 9% of D0 concentrations. On D12, measured residues on leaves were 5.9 times residues on under story vegetation.

Dissipation of diazinon from vegetation in the WA orchards was similar to dissipation from PA orchards. Diazinon residues from apple leaves on D0 in the WA orchards were 5.6 times the residues found on under story vegetation. This difference increased to a factor of 12 by D12. By D4, diazinon concentrations from all sample types were 19% to 25% of D0 concentrations, and 4% to 9% remained by D12.

Diazinon dissipation from each type of vegetation was averaged for individual orchards in PA and WA (Figures 3 and 4). These data demonstrate the variance in residue occurrence that can be encountered during carefully implemented field studies. Dissipation from each collection site within each orchard from the respective states was fit to over 50 polynomial and exponential curves. The best regression fit and power were found for a simple exponential function (Table 3). These fits describe 87-94% of the variance in the diazinon concentrations. Diazinon from leaves dissipated somewhat slower than from grass at both sites. Dissipation equations also predict that residual diazinon present on apple leaves and under story vegetation at D14 should be 6% and 2% of the respective diazinon present at D0.

Earthworms: All earthworms collected from orchards in PA and half of the earthworms from WA contained residues of diazinon (Table 4). Earthworm exposure was log-normally distributed with the highest two concentrations being 163 μg/g for a live collection on A1D12 and 69 μg/g for a worm found dead on the surface A2D2. The highest concentration represents the 96th centile of diazinon measured in PA and the 97th centile among concentrations found in earthworms from all sites. The mean diazinon concentration in earthworms was 10.5±6.02 μg/g. Using log transformed data, worms found alive and dead in PA orchards were found to contain statistically similar diazinon concentrations (p>0.9).

Mean diazinon concentrations in earthworms from WA were 0.39±0.21. Comparisons of log transformed diazinon concentrations in earthworms demonstrated higher exposures in PA orchards than in WA orchards (p<0.005). This difference was also found when comparing live captured earthworms from PA and WA (p<0.017).

Starling GI Tracts: GI tracts from European starling (*Sturnus vulgarus*) nestling carcasses in PA were analyzed for diazinon residues (Table 5). No GI

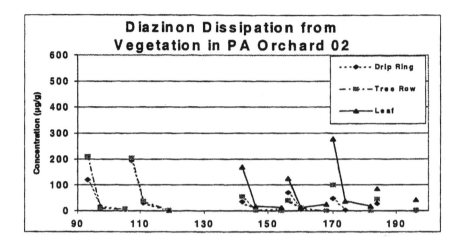

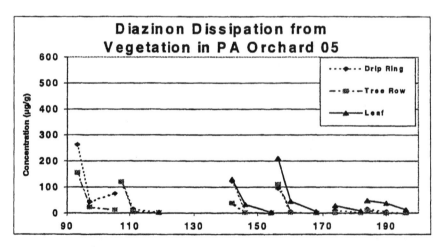

Figure 3. Mean diazinon dissipation from vegetation collected from individual orchards in Pennsylvania following five applications spanning Julian Dates 93-187.

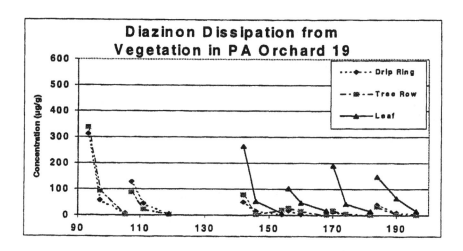

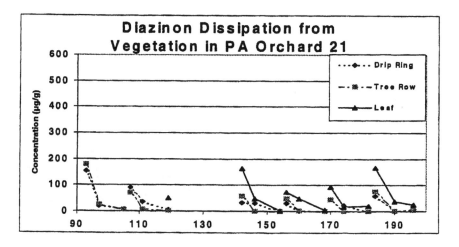

Figure 3. *Continued.*

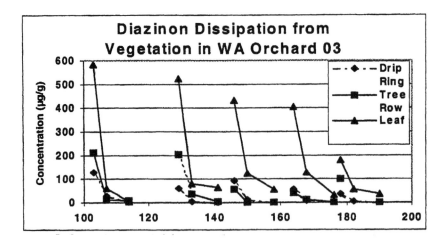

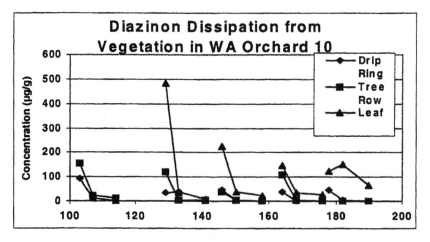

Figure 4. Mean diazinon dissipation from vegetation collected from individual orchards in Washington following five applications spanning Julian Dates 102-178.

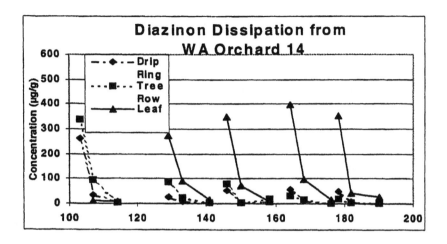

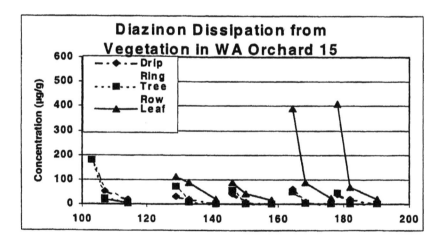

Figure 4. *Continued.*

Table 3. Dissipation of diazinon from vegetation in orchard ecosystems of Washington and Pennsylvania.

State	Matrix	Rate Constant[a] (k)	r^2 [a]	Range of rate constants[b]	
				Low	High
Washington	Leaves	0.20	0.90	0.0599	0.440
	Ground Cover				
	Drip Ring	0.26	0.94	0.130	0.360
	Tree Row	0.27	0.93	0.138	0.389
Pennsylvania	Leaves	0.20	0.87	0.0376	0.565
	Ground Cover				
	Drip Ring	0.27	0.92	0.172	0.386
	Tree Row	0.28	0.82	0.101	0.354

[a] For the expression $C_t = C_0 \exp(-k\,t)$

[b] - Data sets for one Leaf and three Drip ring measurements contained only 2 data points and are not reported in this range of rate constants.

Table 4. Diazinon residues in earthworm samples from apple orchards in Pennsylvania and Washington.

Site	N	Diazinon (µg/g)
PA02	1	69.12
PA05	8	1.203-163.0
PA19	10	0.369-6.703
PA21	9	0.557-3.616
PA Mean	28	10.54±31.8
WA10	6	<0.007-1.612
WA14	2	<0.007- 0.047
WA Mean	8	0.39±0.61

Table 5. Diazinon concentrations (μg/g) from GI tracts of birds collected in apple orchards and control sites from Pennsylvania and Washington.

Sample Types	Type	Application (Mean±sd)	Concentration	n
PENNSYLVANIA				
American Robin	Treatment	0.458 ± 0.704		17
	Control	<0.007		1
Br. Headed Cowbird	Treatment	0.012 ±0.018		7
	Control	<0.007		1
Blue Jay	Treatment	0.016		2
Common Grackle	Treatment	<0.007		3
European Starling	Treatment	0.018 ±0.036		72
	Control Mean	<0.007	37	
Mourning Dove	Treatment	0.025±0.024		7
Northern Cardinal	Treatment	0.006 ±0.006		18
	Control Mean	0.001		6
WASHINGTON				
American Robin	Treatment	0.102 ±0.114		18
European Starling	Treatment	0.011±0.016		132
	Control	<0.007		22
Canada Goose	Treatment	2.17±0.38		12
Killdeer	Treatment	1.82 ± 0.63		4
Mourning Dove	Treatment	<0.007	3	
Black-billed Magpie	Treatment	0.065±0.088		2
Br. Headed Cowbird	Treatment	0.008±0.012		4
W. Meadowlark	Treatment	0.123±0.141		4

tracts from control sites contained diazinon. GI tracts collected from treated orchards contained a distribution of diazinon residues ranging from <0.007 µg/g to 1.044 µg/g. In WA, one of 22 control samples had a trace of diazinon, and the remainder contained no diazinon (Table 5). GI-tracts from treated orchards contained diazinon concentrations ranging from 0.007 to 0.226 µg/g. Distributions of diazinon content in starling GI tracts were heavily skewed toward low concentrations.

Carcass Searched Birds: GI tracts were removed from 81 bird carcasses found on treated orchards in PA and from 70 bird carcasses collected in WA orchards. Distributions of diazinon in these GI tracts were skewed toward low concentrations, and the overall means were 0.092 µg/g and 0.212 µg/g respectively. Diazinon concentrations in GI tracts of 12 Canada geese (*Branta canadensis*) found dead on study sites were elevated compared to all other avian carcasses. GI tracts of these geese contained 2.17± 0.38 µg/g diazinon (19).

DISCUSSION

Diazinon Applications: Wildlife exposures to diazinon can be discussed with some certainty since diazinon was quantified in tank samples and spray rates were monitored closely during pesticide application. Mean residue concentrations within 9% of nominal concentrations may be sufficient evidence of appropriate application, but the amounts of chemical and water added were also measured along with spray output and tractor ground speed. These additional measurements estimated application within 3% of nominal concentrations. Thus we are confident that mean diazinon applications were 91% to 97% of nominal and that the CL95 for applications was 82% to 104% of nominal. The data generated from the study demonstrate the distribution of diazinon concentrations applied to study sites. The majority of samples were near the mean, but significant excursions were observed (Figures 1&2). Such data are essential when designing input distributions for probablistic risk assessments. Evaluations that assume mean concentrations will not represent reality on most fields and will not capture the hazard to non-target organisms that may inhabit portions of orchards that have received higher than average application. Also most estimates for application rates are hypothetical and contain little information regarding the shape of the actual distribution.

Vegetation: Diazinon concentrations were highest in the apple leaves, which is consistent with applications targeted toward tree canopies, where insect control was desired. Once applied, the average rate of diazinon dissipation was consistent among sample types (Table 3). The rate constants (k) demonstrate more rapid diazinon dissipation from under story grass surfaces than from leaf surfaces. These loss rates suggest that significant amounts of diazinon were not washed from leaves into grass by rainfall. Data show the presence of diazinon in all areas of the apple orchards, implying high potential for exposure of non-target species inhabiting these orchards. Furthermore, regression analyses indicate that 2% to 6% of the applied diazinon would be present two weeks following application, at the time of the next application. These estimates are comparable to the 0.7% to 20% of diazinon remaining on individual fields 12 days following application. Therefore, the potential for exposure to target and nontarget species was uninterrupted during the study period.

Even though average degradation rates were quite consistent within a given field, inter-field rates varied more. The distribution of degradation rates observed for each vegetation type in each field shows pseudo first order degradation rate constants (k) that vary from 0.14 day^{-1} to 0.30 day^{-1} with the majority of rate constants falling within 0.02 day^{-1} of the mean (Table 3). This information provides a useful basis of measured diazinon degradation that can be used to parameterize probablistic risk assessments (20). This and similar data should be tabulated to form the basis of exposure assessment in ecological risk assessments.

Earthworms: Consistently high diazinon concentrations were found in earthworms. Organophosphorus (OP) insecticides are toxic to earthworms (21,21). Some OPs may cause earthworm death or perhaps avoidance of OP treated areas. However, the high diazinon concentrations found in live earthworms highlights the potential for secondary poisoning of verminivores in orchard ecosystems. Earthworms are common items in the diet of many birds and are consumed by various other vertebrates (23-25). Given the diazinon concentrations found in earthworms and the relative number of earthworms available for analysis from the two areas, risks that birds would ingest OP containing earthworms is significantly greater in the PA orchards than in those from WA. Diazinon uptake by earthworms in WA was similar to that found in other field studies (26), but concentrations found in earthworms from PA were much higher.

Birds: Diazinon presence in GI tracts verifies avian exposure to diazinon following D-z-n 50W applications of 3.4 kg AI/ ha. Diazinon was present in GI tracts collected from 24 of 25 avian species evaluated in treatment orchards. Quantifiable residues were present in 19 avian species. Residues found in the GI tracts of the 12 Canada geese and four nestling killdeer were the highest noted in this study. The susceptibility of grass-eating waterfowl species to diazinon has been documented previously (6,9). This is largely due to their foraging behaviors.

A worst case scenario for diazinon toxicity to free ranging birds would involve the LD50 for the red-wing black bird, 2.6 mg/kg. This is a conservative estimate of diazinon toxicity to free ranging birds. Red-wing blackbirds are small, 37-65 g (27). Therefore, LD50s can be achieved with as little as 96 µg diazinon. Given a conservative LD50 for avian species and the fact that 90% of an OP insecticide may be excreted by birds that ultimately succumb from OP exposure (16,17), our field data suggest passerine poisoning during the first four days following diazinon application in orchard systems and a substantially decreased risk when exposure occurs after this four day window. During the first four days following diazinon application in the orchards, 78-84% of residues dissipated from vegetation, leaving <23% to adversely impact wildlife thereafter.

CONCLUSIONS

The information presented here demonstrates the variance of diazinon distribution found in apple orchards undergoing intensive pest management. Diazinon dissipation from vegetation follows pseudo first order rates, and dissipation is similar in the two regions. Knowledge of temporal pesticide occurrence patterns in orchards is critical in probablistic assessments of wildlife exposure to insecticides in orchard systems. Such data sets are also valuable as weight of evidence for reasonable estimation of diazinon exposures in other ecosystems.

ACKNOWLEDGMENTS

Funding for this study was provided by Ciba-Geigy Corporation. This study could not have been completed without the efforts of Ron Mellot, Rick Lestina, Chuck Crabtree, as well as analytical and ecotoxicology team members who are too numerous to name individually. Land owner cooperation was also essential in conducting this study. Their efforts are gratefully acknowledged.

Literature Cited

1. United States Environmental Protection Agency. Pesticide Fact Handbook. Vol. 1. Noyes Data Corporation: Park Ridge, N.J. 1988.
2. Eisler R. Diazinon hazards to fish, wildlife, and invertebrates: A synoptic review. NTIS Order No.: PB86-235074/GAR, 1986.

3. Egyed MN, Makinson M., Eilat A., Shlosberg, A. *Refuch. Veterin.*, 1974, 31:22-26.
4. Schafer E.W., Bowles W.A.J. , Hurlbut J. *Arch Environ. Contam. Toxicol.* 1983, 12:355-382.
5. Balcomb R., Stevens R, Bowell C. *Bull. Environ. Contam. Toxicol.*, 1984, 302-307.
6. Kendall R.J., Brewer L.W. *J. Wildl. Dis.*, 1992. 28:263-267.
7. Kendall RJ, Brewer LW , Hitchcock RR. *J. of Wildl. Dis.* 1993, 29:458-464.
8. Mineau P, Boersma DC, Collins B. *Ecotoxicol. Environ. Saf.* 1994, 29:304-329.
9. Stone WB, Gardoni PB. *Northeast Environ. Sci.*, 1985, 4(1):30-38.
10. Hill EF, Camardese MB. *Ecotoxicol Environ Saf.*, 1984, 8 (6): 551-63.
11. Grue CE, Shipley BK. *Stud. Avian Biol.*, 1981, 6:292-296.
12. Grue CE , Shipley BK. *Environ. Res.*, 1984, 35:454-465.
13. Tank SL, Brewer LW, Hooper MJ, Cobb GP. *Environ. Toxicol. Chem.*, 1992, 12:2113-2120.
14. Robinson SC, Kendall RJ, Robinson R, Driver CJ, Lacher TE. *Environ. Toxicol. Chem.*, 1988, 7:343-349.
15. Hooper, MJ, Dietrich PJ, Weisskopf CP, Wilson BW. *Bull. Environ. Contam. Toxicol.*, 1989, 42:651-659.
16. Cobb GP, Hooper MJ. Nonlethal wildlife monitoring to determine exposure to xenobiotics and resulting impacts. in RJ Kendall and TE Lacher, eds. The Population Ecology and Wildlife Toxicology of Agricultural Pesticide Use, Lewis Publishers, Chelsea, MI. 1994.
17. Kendall, RJ, Akerman J. *Environ. Toxicol. Chem.*, 1992, 11:1727-1749.
18. Brewer, LW, Hummell RA. What is measurable in wildlife toxicology. in RJ Kendall and TE Lacher, eds., Wildlife Toxicology and Population Modeling. Lewis Publishers :Boca Raton, FL. 1994.
19. Cobb GP, Bens CM and Kendall RJ , Mellott RJ, Brewer LW. *Environ. Tox. Chem.* 2000, 19(5):1360-1367.
20. USEPA. ECOFRAM: Terestrial Report. United States Environmental Protection Agency, Environmental Fate and Effects Division. Washington, DC. 1999.
21. Roberts BL, Dorough HW. *Environ. Toxicol. Chem.*, 1984,, 3:67-78.
22. Cobb GP, Hol EH, Allen PW, Gagne JA, Kendall RJ. *Environ. Toxicol. Chem.* 1995, 14:279-285.
23. Webster W.D., Parnell J.F., Biggs, WC. Mammals of the Carolinas, Virginia, and Maryland. Univ. North Carolina Press, Chapel Hill, 1985.

24. Martoff B.S., Palmer W.M., Bailey J.R., Harrison J.R. Amphibians and reptiles of the Carolinas and Virginia. Univ. North Carolina Press, Chapel Hill, 1980.

25. Conant R. A field guide to reptiles and amphibians of eastern and central North America. Houghton Mifflin Co., Boston, MA., 1986.

26. Stephenson G. L., Wren C. D., Middelraad I. C. J., Warner, J. E. *Soil Biol. Biochem.*, 1997, 29: 717-720.

27. Terres JK, The Audobon Society Encyclopedia of North American Birds. Alfread A Knopf, Inc: New York, NY. 1980.

Chapter 13

Field Studies of Imidacloprid Distribution Following Application to Soil through a Drip Irrigation System

Allan S. Felsot[1], Robert G. Evans[2], and Judy R. Ruppert[1]

[1]Department of Entomology, Food and Environmental Quality Lab,
Washington State University, Richland, WA 99352
[2]Northern Plains Agricultural Research Laboratory, Agricultural Research
Service, U.S. Department of Agriculture, Sydney, MT 59270

Both surface and subsurface drip irrigation can reduce overland flow and thereby reduce surface transport of pesticides. Little is known, however, about leaching of pesticides when applied via drip systems. A series of experiments were conducted over several years to characterize the horizontal and vertical distribution of imidacloprid [1-[(6-chloro-3-pyridinyl)methyl]-N-nitro-2-imidazolidinimine] in an experimental hop yard under subsurface drip irrigation. The insecticide was applied once by injection directly into the irrigation pipe buried 45-60 cm deep on each side of a hop vine row. Water exited the pipe through labyrinth emitters spaced at 90-cm intervals. Soil profiles of either 105 cm or 150 cm in depth were collected at various times following chemigation to characterize imidacloprid distribution in successive 15-cm soil layers. Imidacloprid residues were mostly confined to soil profiles collected within a 30-cm radial distance from an emitter. When irrigation occurred on a 4-h daily time schedule (treatment 4H), imidacloprid leached to the lowest depths sampled (105 or 150 cm). Imidacloprid leaching was comparatively less extensive in two of the three trials that studied irrigation scheduling triggered on and off by soil moisture sensors recognizing pre-defined soil matric potentials (irrigation treatment MP). Pseudo-first order half-lives of imidacloprid ranged from 18 d in 4H treatments to 31 d in MP treatments. All dissipation rates were faster than previously reported for field studies at other locations. The effect of analyzing duplicate profile samples individually rather than analyzing them as bulked composites was investigated in a randomized treatment design experiment. Individual analyses of replicate plot duplicates significantly lowered the overall treatment mean standard deviation, suggesting an improved potential for resolving differences in residue distributions that may result from changes in irrigation management practices.

To improve efficiency of water use and reduce soil erosion, growers in semi-arid regions are increasingly foregoing furrow systems in favor of surface (DI) and subsurface drip irrigation (SDI). Drip irrigation is the application of water through emitters spaced at defined intervals in pipes that are placed either on the soil surface or buried below the surface. In California and the Pacific Northwest (PNW), drip irrigation is used on about 0.2 million hectares. About 20,000 hectares are drip irrigated in Washington State (*1*). DI and SDI are commonly used in perennial crops, e.g., orchards, vineyards, and hop yards, but many agronomic crops (e.g., corn, cotton) and horticultural crops (e.g., tomato, asparagus) also use drip irrigation (*2*).

The growth of interest in SDI is commensurate with its purported advantages which include: increased water use efficiency, enhanced plant growth and yield, reduced salinity hazard, decreased energy requirements, improved cultural practices, limited weed growth, elimination of surface runoff, and improved application of fertilizer and other chemicals (*2,3*).

Using irrigation systems to apply fertilizers and pesticides is commonly referred to as chemigation. Chemigation via DI and SDI (i.e., drip chemigation) also has advantages compatible with environmental stewardship: no worker exposure to foliar pesticide residues, reduction of waste from cleaning out spray tanks, elimination of drift, and less exposure of biological control organisms to pesticides in integrated pest management programs. Systemic pesticides seem particularly well suited for application by drip chemigation and would be compatible with IPM if plant uptake was rapid enough to allow delay in application until a pest problem was developing.

Despite the advantages of drip irrigation systems, further expansion of drip irrigation technology has been limited by several concerns: costs of implementation, difficulty of laying pipe in certain soil types, management of blocked emitters, and the lack of enough data to convince growers that it offers a superior method of fertilizer and pesticide application.

Particularly needing solutions are questions about pesticide uptake by plants and potential leaching in soil. While the efficacy of chemigated systemic insecticides seems adequate, especially for sucking insect control, plant uptake kinetics will determine if applications can be made in an emergency rather than made prophylactically. Also, residues in harvested commodities should be substantially less than the tolerance levels, and preferably minimized to reduce dietary exposure.

The effects of DI and SDI on leaching of pesticides injected into water and emitted as point sources into the soil, as opposed to leaching by water after application directly to the soil, has hardly been studied, especially under field conditions. Such leaching concerns are prompted by studies showing chemicals applied in flood irrigation water can leach to greater depths than chemicals applied directly to the soil surface (4). Chemicals have also been observed to rapidly leach along preferential flow paths when applied with irrigation water (5). Studies should focus on the initial distribution of chemigated pesticides and the translocation of residues as the growing season progresses.

In addressing concerns about pesticide leaching, we monitored the distribution of imidacloprid aphicide in a hop yard following injection into an SDI system. (6). Imidacloprid is a comparatively low toxicity systemic insecticide with both foliar and soil bioactivity, but its high water solubility (500 ppm) and low soil distribution coefficients (7) suggest that it may be comparatively mobile. In our early studies, irrigation was supplied on a daily 4-h schedule (6). Imidacloprid residues were observed below the emitters at the maximum sampling depth of 105 cm, suggesting that fixed-schedule water delivery might promote leaching.

Observations from our earlier study led to the hypothesis that vertical movement of imidacloprid could be reduced by irrigating only when soil water content fell below an optimum matric potential. Over several growing seasons, we have conducted various field experiments to help test our ideas about best management practices for chemigation via drip systems. In an effort to further product understanding, we now report the results from several field experiments designed to determine horizontal movement around the irrigation emitters, vertical leaching under two different irrigation scheduling regimes, and the effect of sampling replication on residue variability.

Materials, Methods, Procedures

Experimental Field Location and Description

The study site was located in an experimental 1-ha subsurface drip-irrigated hop yard that was developed in 1992 at the Washington State University Irrigated Agriculture Research and Extension Center (IAREC) near Prosser, WA. The station lies in the eastern end of the Yakima Valley and receives an annual rainfall of 15-25 cm. The soil was classified as a Warden very fine sandy loam (coarse-silty mixed mesic Xerollic Camborthids), which is common in the lower end of the Yakima Valley. A composite sample from the top 60 cm had a 49% silt and 3.2% clay content, a 1.03 % organic matter (OM) content, and a pH of 8.0. Soil collected between 60 cm and 105 cm had a pH of 7.8, an OM content of 0.74 %, and a silt and clay content of 48% and 5.2%, respectively. Gravimetric soil water content (w/w) at 33 kPa (moisture holding capacity, MHC) in the top 60 and lower 60 cm of the sampled profile was 18.7 and 18.1%, respectively.

The yard was planted with four varieties of hops (Chinook, Willamette, Columbus, and Mt. Hood) laid out in four 0.2-ha blocks (Figure 1). Each block was divided into 12 separate plots, each with 5 rows spaced approximately 2.1 m apart and consisting of 7 hop vines in hills, also spaced 2.1 m apart. Each row was irrigated via one subsurface polyethylene pipe (1.32-cm ID) buried approximately 45-60 cm deep. Each pipe ran parallel to the hop rows and was placed within 60 cm of the north side of each row. The pipe was fitted with labyrinth emitters spaced 90 cm apart. The labyrinth consisted of a small plastic insert molded into a series of circuitous channels in which water moving down the central pipe opening could enter and exit through a tiny hole to the outside of the pipe. Water was emitted at a rate of 1.9 L/h (~1 mm/h).

Irrigation Scheduling

Each of the 12 plots within a variety block could be independently irrigated and chemigated (Figure 1). A 12-valve manifold sat at the top of the yard. Water was pumped into the manifold from a well. Each of the 12 lines contained an injector port for application of pesticides or special fertilizer treatments. Normally, all the blocks were fertigated uniformly (140 kg/ha N; 28 kg/ha P). Irrigation scheduling was controlled either through a timer or through feedback from buried soil water sensors installed in the center of each variety block (1996-1998) or in individual plots (1999) (Figure 1). Two types of sensors — frequency domain reflectometers (FDR, Campbell Scientific 615L) and standard tensiometers—were buried at a depth of ~40 cm and connected to a datalogger (Figure 1). During 1996, water was automatically turned on for 4 hours daily (treatment **4H**). During 1997-1999, water turned on when soil matric potential dropped to -15.5 kPa and turned off when matric potential reached -14.5 kPa (treatment **MP**).

Horizontal and Vertical Distribution of Imidacloprid—Fall 1996

We previously reported observations of the horizontal and vertical distribution of imidacloprid within a 30-cm radius of the emitters during June through August of 1996 in variety block Willamette (*6*). In September 1996, just prior to harvest of the hop cones but when the plants were going into senescence, we repeated the study on a previously untreated plot of hop variety Willamette. Distribution of imidacloprid was monitored for an eight day period

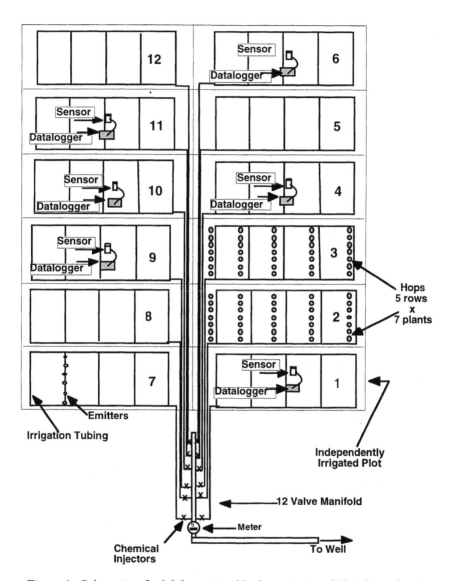

Figure 1. Schematic of a 0.2-ha variety block consisting of 12 independently irrigated plots in the experimental drip irrigation hop yard at WSU-Prosser.

to confirm earlier observations and determine possible horizontal movement of imidacloprid beyond 30 cm.

Prior to application, the emitters were located by exposing the pipe at each end of the plot. A field tape measure was laid as a transect between the exposed emitters, and the locations of the remaining emitters were located at 90-cm intervals along the transect and delineated with flags for later sampling.

After the emitters were located, imidacloprid (formulated as Provado) was mixed in water and injected by a peristaltic pump into the irrigation system through a manifold that led to separate irrigation lines for each plot within a variety block (Figure 1). The lines were flushed for 30 minutes with water following insecticide injection. The rate of application was 0.28 kg AI/ha.

To determine the horizontal and vertical distribution of imidacloprid, a series of soil samples were taken to a depth of 105 cm at various distances from the emitter in a pie-shaped distribution (Figure 2). We also took profile samples at a 45-cm distance from the emitter along the direction of the pipe and in a 45-cm distance perpendicular to the emitter in the direction of the interrow area. Nine profiles were collected on days 1, 3, and 8 after application.

Soil samples at each numbered location (Figure 2) were collected in successive 15-cm depths using a bucket auger (5-cm diameter). The top and bottom 2.5 cm of each depth were removed to avoid cross contamination between layers and from surface soil that may have fallen into the bore hole.

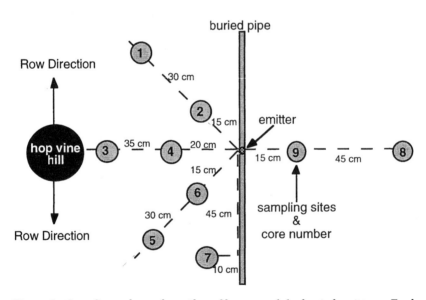

Figure 2. Sampling scheme for soil profiles around the buried emitters. Each numbered circle represents a soil core taken to a depth of 105 cm. Distances along the dashed transect lines represent the length between the marked location of the emitter and the center of each profile.

Between samples the bucket auger was brushed in water and then dried before taking the next sample. Samples were held in double polyethylene bags, returned to the lab, and stored at –20 °C until analyzed.

During July and August of 1997, approximately 10 months after application, replicate soil cores were collected from the plot treated the previous September. A single profile was sampled within 15 cm of the emitter using a 10-cm diameter bucket auger. The soil was again sampled in 15-cm increments, discarding the upper and lower 2.5-cm of soil. Soils were held in storage until analysis as previously described.

Vertical Distribution of Imidacloprid with Irrigation Under Matric Potential Control (1997-1999 Field Studies)

During 1997 and 1998, variety Willamette plots that had been treated during the early summer of 1996 were retreated with 0.28 and 0.22 kg imidacloprid/ha, respectively. Insecticide application occurred after the emitters were located as described for the September 1996 experiment. Irrigation water was turned on automatically whenever soil matric potential dropped below –15.5 kPa. A single FDR was buried at a depth of 40 cm in the center of the variety block to control the irrigation scheduling and shut off the water when the matric potential rose to –14.5 kPa.

Soils were sampled just prior to application and seven times later over the next 70 days. Each collection consisted of two randomly chosen locations, one each in rows two and four of the plots. The profile was started within 15 cm of the emitter on the hop row side. Soil was collected with a 10-cm diameter bucket auger in 15-cm layers to a depth of 150 cm. Soils were handled as previously described.

During 1999, studies were switched to variety block Chinook using plots that had not been treated with imidacloprid for at least two growing seasons. Replicate plots were delineated to compare two irrigation schedules simultaneously—4H and MP. FDR instrumentation and tensiometers were moved to the center of each replicate treatment plot (Figure 1). After application of imidacloprid (0.22 kg/ha), soil was collected several times over the next 2 months and handled as described for the 1997-98 experiments. Two randomly chosen profiles were collected from the second and fourth row of each of the replicate treatment plots, making a total of four profiles sampled per treatment. Within two weeks of application, however, it was discovered that one of the MP replicate plots had not actually received an insecticide application. At this point an extra profile was collected from its corresponding replicate on the remaining sampling days.

Sampling Variability Experiment

During June 2000, imidacloprid distribution experiments were moved to variety block Mt. Hood using plots that had not been treated during the previous two growing seasons. Soil moisture monitoring instrumentation was set up as previously described for the 1999 experiments. To achieve the objective of better quantifying imidacloprid distribution following application and characterizing sampling variability, three replicate plots were each delineated for irrigation treatment 4H or MP. Following imidacloprid injection (0.22 kg/ha), soil samples were collected within 24 h and 7 d later.

Analytical Methods

During 1996-1998, soils were mixed by hand prior to removing a sample for analysis. Owing to the low water content of the soils and the lack of structure (due to low clay and organic carbon content), it was not necessary to sieve the soils. To characterize the variability of laboratory subsampling, six 20-g subsamples were removed from each of five previously collected field samples having average concentrations ranging from 2-640 ng g^{-1} (ppb) The coefficient of variation (CV) for recovered residues ranged from 12%-39% (mean = 25%). To minimize variation in laboratory subsampling further, soils collected during 1999 and 2000 were laid flat on paper, mixed back and forth, and then approximately one-quarter of the total was removed for laboratory analysis and long-term storage ($\sim$-20°C). The more extensive mixing did not appreciably reduce the CV for subsampling.

Soils were thawed for 24 hours at $\sim$6°C for 24 h prior to analysis. Ten grams of soil were removed to determine % moisture (w/w). Thirty grams of soil were weighed into centrifuge bottles and extracted twice by reciprocal shaking with either 50 mL of water (1996 samples as described in 6) or a 9:1 mixture of acetonitrile (ACN) and deionized water (1997-1999). In 2000, 0.01M CaCl$_2$ was substituted for water. After each extraction, samples were centrifuged for 10 minutes and the supernatant was filtered under vacuum through Whatman 934H glass microfiber filters. The acetonitrile was removed by vacuum rotary evaporation. The remaining water from the 1996-97 extracts was partitioned twice with 75 mL of CH$_2$Cl$_2$ and then further processed as described in 6. The resulting water extracts for the 1998-2000 samples were passed through a Bakerbond C18 SPE 6-mL cartridge. Imidacloprid was eluted from the cartridges with two 4-mL aliquots of ACN. The ACN was evaporated to dryness under nitrogen in a water bath (60 °C), and then the extract was reconstituted in 1 mL of 1:1 ACN:water. All finished extracts were passed through a 0.45-μm Acrodisc filter and stored in brown 1.8-mL vials for HPLC.

Imidacloprid was detected and quantified on a Varian HPLC using a 4.5-mm x 15-cm C18 reversed phase column. Imidacloprid was eluted using a program gradient of 95:5 ACN:H$_2$O to 100% ACN over a 20-min period. Imidacloprid was detected with a photodiode array detector set to monitor absorbance at 268 nm. The spectral analysis mode was used as necessary to qualify questionable peaks eluting at the expected retention time for imidacloprid. Based on the ability of the detector to produce a reliable signal above background when 60 μL of a 0.03 μg mL^{-1} calibration standard was injected, the method detection limit was set at 1 ng g^{-1} (1 ppb) oven dry soil.

Every day that field samples were extracted, one or two aliquots of soil collected from an untreated plot were freshly amended with imidacloprid to yield a concentration of 50 ppb. Extraction efficiency over the four years of the reported studies was 90% or greater with a CV less than 15%.

Results and Discussion

Horizontal and Vertical Distribution of Imidacloprid—Fall 1996

One problem we encountered was determining where to sample a soil profile relative to the location of the emitter. Because published observations of water patterns around drip emitters showed a radial distribution with decreasing water potential as distance from the emitter increased (8,9), we hypothesized that imidacloprid would stay comparatively close to the emitter along any horizontal direction. We also assumed that imidacloprid distribution around the emitter would be homogeneous so that a sample taken on any side of the pipe would be representative. During the 1996 summer studies, imidacloprid was found within 7 days of application to be distributed in all five cores taken within a 30-cm radius of the emitters (6). Cores taken closest to the emitter did not necessarily have the highest concentrations of imidacloprid.

During the fall 1996 experiment, imidacloprid was generally found in the injection zone (45-60 cm) at the highest concentrations in the cores within 15-20 cm of the emitter (Figure 3, cores 2, 4, 6 on day 1). This pattern was also exhibited on day 3. By day 8, imidacloprid concentrations had significantly declined in all cores but continued to be detected at a radial distance of 30 cm. Imidacloprid was never detected in core 9 that was located 45 cm from the emitter on the north side of the irrigation pipe.

Imidacloprid residues were not detected below 75 cm the day after application, but >1000 ppb was detected in core 4 at the lowest sampling depth (105 cm) on day 3 (Figure 3). By day 8, imidacloprid was detected throughout the profile, but its rapid dissipation suggested leaching below the sampling zone. Soil moisture (data not shown) at the lower depths did exceed the moisture holding capacity in some cores, suggesting saturated flow and consequently, enhanced mobility. During the fall, the hop plants are unlikely to need much water, so a daily four hour irrigation was likely excessive.

To test the hypothesis of leaching below the sampling zone, during July and August of 1997 replicate cores were sampled to a depth of 150 cm at a distance of 15 cm from the emitter. Replicate soil samples were extracted with either water (as described in 6) or with 9:1 ACN:water. Imidacloprid residues were observed in cores collected 10 months after application at a depth of 120 cm (Figure 4). Although residues had plenty of time to "age", both water and ACN extracted similar amounts of imidacloprid.

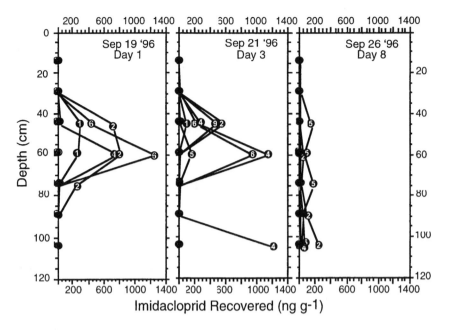

Figure 3. Recovery of imidacloprid residues in cores within a 45 cm radius of a drip irrigation emitter following application on September, 18, 1996. Refer to Figure 2 for location of core numbers relative to emitter position.

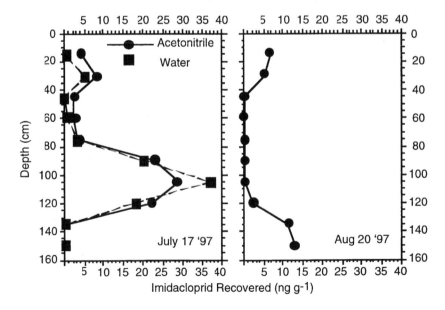

Figure 4. Distribution of imidacloprid residues in soil treated during Sept. '96.

Other research has suggested that with increasing incubation times in soil, the soil distribution coefficient of imidacloprid increases, making water a comparatively poorer extracting solvent (10,11). If aged residues are less available to extraction by water, than soil mobility may also be allayed. While our past work also suggests an aging effect for field-collected imidacloprid residues (6), water was still an efficient extracting solvent. ACN extraction of duplicate cores collected 11 months after application indicated further leaching of "aged" residues to a depth of 150 cm (Figure 4). The residues observed in the uppermost layers of the profile may represent translocation in water that sometimes wetted the soil surface, presumably when the soil became saturated in the vicinity of the emitters.

Vertical Distribution of Imidacloprid with Irrigation Under Soil Matric Potential Control (1997-1999 Field Studies)

The deep leaching of imidacloprid significantly below the emitter zone with daily 4-h irrigation led to an examination of alternative irrigation scheduling as a best management practice. Many growers automatically irrigate using a set schedule each day during the hottest periods of the summer, which occur during July and early August in south central Washington. However, growers are increasingly interested in matching water utilization with crop needs. Thus, if the optimal moisture content for crop growth was determined, then water could be turned on and off in pulses to keep the soil moisture in this range. The problem with deployment of this strategy is determining the most efficient placement and appropriate number of soil moisture sensors.

During 1997 and 1998, we determined the distribution of imidacloprid in soil cores with irrigation controlled by feedback from soil moisture sensors located in a central position in an entire 0.2 ha block. During 1997 (Figure 5), imidacloprid remained in its injection zone one week after application, moved to a depth of 105 cm (22 ppb recovered) within 23 days, but dissipated from this depth by day 43 and day 75 (data not shown) after application. During 1998 (Figure 5), in contrast, imidacloprid was recovered at 90 cm on day 21 (95 ppb) and on day 42 (65 ppb). At 75 days in both years, imidacloprid was found in the deepest soil layers, but residues were also found at this depth in pre-application samples (Figure 5, Day -1). The plot had been treated for at least four years in a row, and residues had not completely dissipated.

Low levels of imidacloprid from previous growing seasons seemed mobile and significant amounts were still extractable in water (data not shown). A published experiment on aphid control in the same plots also showed that the low ppb levels of imidacloprid may still be bioavailable (12). For example, significant aphid control occurred in insecticide-untreated plots during 1997 if the plots had been chemigated with imidacloprid either one or two years before. Untreated plots without past imidacloprid injections had significantly more aphids (12).

During 1999, we monitored imidacloprid distribution under both 4H and MP irrigation schedules simultaneously within a replicated plot design. To

200

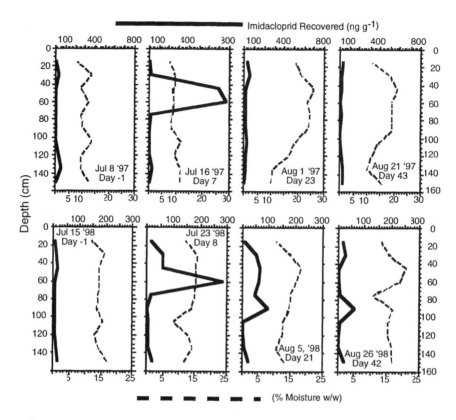

Figure 5. Soil profile distributions of imidacloprid under irrigation controlled by soil matric potential during 1997 and 1998.

avoid the confounding factor of imidacloprid residues from a previous years application, experiments were moved to variety block Chinook in 1999. Plots were chosen for study if imidacloprid had not been used for at least two or more years. Soil moisture sensors were placed within the center of each replicate treatment plot.

Recoveries of imidacloprid were similar 24 hours after application (~700 ppb) (Figure 6). By day 15, imidacloprid remained around the injection zone in the MP treatment but was detected at the 90-cm depth in the 4H treatment. By day 28, some wetting at the soil surface was noted and imidacloprid was recovered in the top 30 cm. After two months, imidacloprid residues in the MP treatment were mostly recovered from the injection zone and above but had moved to the 120-cm depth in the 4H treatment.

During the first month after application, gravimetric moisture content analysis suggested that soils in the emitter zone of both irrigation treatments were near or above MHC (Figure 6). Pertinently, water pulsed on and off in the MP treatment, possibly allowing some drying to occur and retardation of

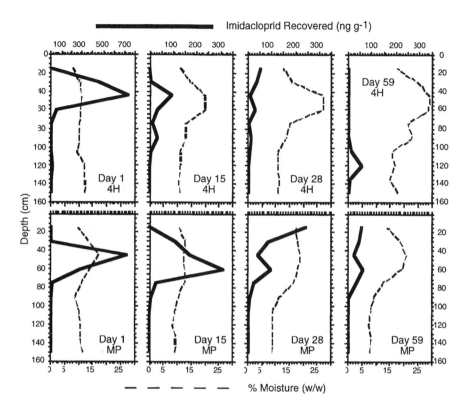

Figure 6. Imidacloprid distribution in soil under two irrigation management schedules, 4-hour daily (4H) and matric potential control (MP).

imidacloprid leaching. At two months post application, moisture content exceeded MHC in the 4H plot at the 90-cm depth, suggesting a greater potential for enhanced leaching of imidacloprid below this level. Coincidentally, at depths ≥90 cm, nearly 20 times more imidacloprid was found in the 4H profiles than was found in the MP profiles (Table 1). However, nominally more imidacloprid was found below 90 cm in the MP profiles during the 1997 and 1998 growing seasons than was found in the 1999 growing season (Table 1). On the other hand, the cumulated concentration in the 1997 MP profiles was still nominally less than in the 1999 4H profile.

Owing to the comparatively high water solubility of imidacloprid, concerns have been expressed about its leaching potential. These concerns have been allayed, however, by laboratory studies showing increases in soil distribution coefficients with decreases in concentration and decreases in mobility with residue aging (7,10,11). Field studies also indicated less mobility than the water solubility would indicate (11). In contrast, we showed that under irrigated conditions with subsurface applications, imidacloprid moved beyond the

Table I. Estimates of First-Order Half-Life for Dissipation of Imidacloprid
and Cumulative Residues (ppb) Recovered at a Depth ≥90 cm

Year	Irrigation Control	First Order $T_{1/2}$ (days)	R^2	Cumulative Residue @ ≥90 cm
1997	Matric Potential	21.9	0.73	15
1998	Matric Potential	29.9	0.77	60
1999	Matric Potential	30.9	0.87	4
1999	4 Hour Daily	17.6	0.73	73

injection zone to depths of 150 cm. However, after drip chemigation from surface emitters in a commercial hop yard, imidacloprid remained in the top 90 cm of the soil profile through harvest (13). As shown in our 1999 experiments, control of irrigation scheduling by soil matric potential monitoring and feedback can also prevent leaching of imidacloprid in subsurface systems.

Another important consideration for determination of leaching potential is residue persistence. To determine how quickly imidacloprid dissipated from the soil profiles, the bulk density and mass of soil collected in each 15-cm layer were assumed to be constant so that concentrations could simply be added. The concentration data were natural log transformed and subjected to linear regression analysis for calculation of a pseudo first order half-life (Table 1). Over the three years of the study, imidacloprid half-life in MP irrigation ranged from 21 to 30 days but was only 18 days in the 4H irrigation treatment. These half-lives are shorter than the 40-42 day half-life reported from experiments in non-irrigated sugar beets (14,15). In the presence of manure fertilizer, the half-life increased to about 100 days, suggesting that sorption may have affected bioavailability for microbial and plant uptake.

Most studies have shown little mineralization of imidacloprid in soil and a comparatively low percentage of metabolite recovery (11,17). Imidacloprid is unlikely to undergo hydrolysis under agronomic pH ranges (16). The influence of plants on pesticide dissipation was clearly shown in another experiment that compared imidacloprid half-life in bare soil and in soil with a ground cover (17). Imidacloprid half-life decreased from 190 days in soil without plants to 45 days in soil with the cover crop. In our hop yards, imidacloprid dissipation may have been enhanced by the tendency of hop roots to grow toward the emitters, increasing the potential for uptake. In other studies we showed that uptake of imidacloprid by hop plants was very rapid after injection into the irrigation system and remained at $\mu g\ g^{-1}$ (ppm) levels in leaves until cone harvest (13).

Sampling Variability Experiment

One limitation to developing best management practices for soil, water, and pesticide management is the need to compare different systems in the field and determine quantitatively their effectiveness. Our studies through 1998 were essentially qualitative in nature; either a 4H system or an MP system was studied in any one year. The 1999 studies attempted to compare systems in the same year. Thus, a replicate plot design was used but individual profiles were treated as independent experimental replicates. Nevertheless, even with three degrees of freedom, the coefficients of variation were too large to discern statistical differences between residues recovered from any depth on any day of sampling. In an attempt to overcome this limitation, sampling replication was increased during the 2000 growing season.

Three replicate 4H and MP plots were delineated in a randomized design. Two randomly selected cores were collected per replicate plot. Normally, duplicate cores are bulked together before analysis in an effort to save resources. We decided to analyze the 45- and 60-cm layers from each core separately, effectively doubling the number of samples to process. A similar sampling strategy was deployed in a previous field study of waste herbicide dissipation in soil following different application treatments (18).

Although the experimental protocol called for allowing the irrigation lines to flush for 30 minutes following application, the lines leading to the MP plots were inadvertently shut off shortly after injection. Thus, samples collected within 24 hours after application failed to contain imidacloprid residues. Samples collected seven days after application, however, did have residues at levels approaching 500 ppb.

The advantage of analyzing individual duplicate cores rather than bulking them together is shown by comparing the standard deviations (SD) of the means for the duplicates with the SD of the treatment means (Figure 7). In all treatments the SD for the treatment means were substantially reduced below the highest SD among the plot duplicate means. For example,1150 ±1532 ppb of imidacloprid was recovered in plot 9 of treatment 4H, Day 1 (Figure 7). The overall treatment mean and SD was 743±491 ppb. If the six cores from the three plot replicates had been handled as six independent replicates rather than as three sets of duplicates, then the imidacloprid concentration would have been 743±833 ppb.

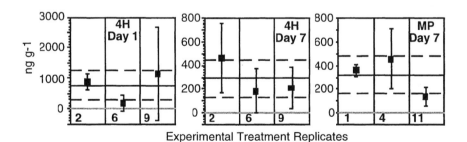

Experimental Treatment Replicates

Figure 7. Mean (squares) and standard deviation (vertical lines) of imidacloprid residues recovered in the combined 45- and 60-cm depths for individual duplicate cores collected in each replicate plot of irrigation treatments 4H and MP. Bold horizontal lines represent the overall treatment means and dashed lines represent the overall treatment standard deviations.

The observed variability in residues among different soil profiles of the same treatment was unlikely to have arisen from laboratory subsampling errors or cross contamination of successive profile soil layers. As previously discussed under 'Analytical Methods', the CV for subsampling averaged 25%, and care was taken in the field to eliminate cross contamination. Thus, the variation between profiles of similar treatments more likely resulted from differences in flow path of emitted water along the length of irrigation pipe and the difficulty in precisely sampling this flow path for solutes. Field sampling variability may be difficult to overcome, but its negative effects on quantitative comparisons between management practices can be alleviated by analyzing individual cores from an experimental replicate rather than bulking them together before analysis.

Conclusions

Because of the need for soil and water conservation in irrigated regions, growers have been gradually adopting drip irrigation systems. Pesticide behavior must be studied in these systems to learn how irrigation scheduling can be manipulated to control agrochemical movement. As a candidate systemic insecticide for chemigation through drip irrigation systems, imidacloprid makes a good model for studying the effects of water management. We showed that imidacloprid can move a radial distance of at least 30 cm from a drip emitter, but the distribution is not homogeneous. Over the course of a growing season, imidacloprid can leach to a depth of at least 150 cm, especially if irrigation is scheduled without attention to the optimal soil matric potential. However, if irrigation scheduling is triggered by a predefined soil matric potential that is matched to crop needs, imidacloprid leaching can be significantly retarded. Field studies that attempt to quantitatively compare different management practices can benefit from individual analyses of duplicate plot samples rather than analyses of composite samples.

Acknowledgements

This research is a contribution from the Washington State University College of Agriculture & Home Economics and was funded in part by grants from the USDA-NRI Program, the Hop Research Council, and the Washington Hop Commission.

References

1. Anonymous. *Irrigation Journal*, **2000**, *January/February*, p.16.
2. Camp, C. R. *Trans. Am Soc. Agric. Engineers* **1998**, *41*,1353.

3. Bucks, D. A.; Davis, S. In *Trickle Irrigation for Crop Production*, Nakayama, F. S. and D. A. Bucks, Eds.; Elsevier Publications: Netherlands, 1986; Chapter 1.

4. Jaynes, D. B.; Rice, R. C.; Hunsaker, D. J.. *Trans. Am. Soc. Agric. Engineers*, **1992**, *35*, 1809.

5. Jaynes, D. B.; Bowman, R. S.; Rice, R. C. *Soil Sci. Soc. Am. J.* **1988**, *52*, 618.

6. Felsot, A. S.; Cone, W.; Yu, J.; Ruppert, J.R. *Bull Environ. Contam. Toxicol.* **1998**, *60*, 363.

7. Cox, L; Koskinen, W. C.; Yen, P. Y. *J. Agric. Food Chem.* **1997**, 45, 1468.

8. Chesness, J. L.; Dryden, J. R.; Brady, U. E., Jr.. *Trans. Am. Soc. Agric. Eng.* **1976**, *19*, 105.

9. Amali, S.; Rolston, D. E.; Fulton, A. E.; Hanson, B. R.; Phene, C. J.; Oster, J. D. *Irrigation Science* **1997**, *17*, 151.

10. Cox, L.; Koskinen, W. C.; Yen, P. Y. *Soil Sci. Soc. Am. J.* **1998**, *62*, 342.

11. Oi, M., *J. Agric. Food Chem.* **1999**, *47*, 327.

12. Wright, L. C.; Cone, W. W.; *J. Agric. Urban Entomol.* **1999**, *16*, 59-64.

13. Felsot, A.; Evans, R. G.; Tallman, L.. In *4th Decennial National Irrigation Symposium*; R. G. Evans; Benham, B. L.; Trooien, T. P., Eds.; American Society of Agricultural Engineers, St. Joseph, MI, 2000; p 416.

14. Rouchaud, J.; Gustin, F.; Wauters, A.. *Bull. Environ. Contam. Toxicol.* **1994** *53*, 344.

15. Rouchaud, J.; Gustin, F.; Wauters, A. *Bull. Environ. Contam. Toxicol.* **1996,** *56*, 29.

16. Zheng, W.; Liu, W. Pestic. Sci. **1999**, *55*, 482.

17. Scholz, K.; Spiteller, M. *Brighton Crop Protection Conference Pests Diseases* **1992**, *2*, 883.

18. Felsot, A. S.; Mitchell; J. K.; Bicki, T. J.; Frank, J. F.; Am. Chem. Soc. Symp. Ser. **1992**, *510*, 244.

Chapter 14

Environmental Fate of Fluometuron in a Mississippi Delta Lake Watershed

Martin A. Locke*, Robert M. Zablotowicz, and Lewis A. Gaston

Southern Weed Science Research Unit, Agricultural Research Service, U.S. Department of Agriculture, P.O. Box 350, Stoneville, MS 38776 *Corresponding author: telephone: 662–686–5272, mlocke@ars.usda.gov

The Mississippi Delta Management Systems Evaluation (MSEA) Project was established to assess management practices for restoring soil and water quality. This paper reviews part of our research consisting of field-scale characterizations of soil properties and their relation to fluometuron degradation and sorption to soil, the role of best management practices (e.g., vegetative strips, riparian zones) in minimizing off-land fluometuron movement, and herbicide concentrations in surface water. Surface (0-5 cm) soil samples were collected in an oxbow lake watershed (Beasley) (60-m spaced grids, 40 ha). Soils were characterized and fluometuron sorption assessed. Based on characterizations, areas representative of different soil series were sampled to evaluate fluometuron dissipation under field conditions. Geostatistical analysis of soil characteristics showed significant spatial dependence, reflecting variability in the alluvial soils. Fluometuron sorption was correlated positively with organic carbon (OC) and clay, and negatively with sand content. Half-lives for fluometuron dissipation from the surface 0 to 5 cm ranged from 12 to 25 days. Fluometuron dissipation was correlated with both clay (positive) and sand content (negative), but not OC. The primary metabolite, desmethyl fluometuron (DMFM) was observed, usually 2 to 3 weeks after herbicide application and occurred in greater concentrations in clay soil. Both fluometuron and DMFM were observed in lake water, appearing one month (June) after field application, and were below detectable limits in October. These studies indicate that fluometuron and its major metabolite have limited persistence in both soil and surface water of the watershed studied.

Understanding the interactions among crop management practices, soil characteristics, and the fate of herbicides is a rational approach for minimizing risks of pesticides on the environment. The EPA considers pesticides one of the major classes of chemical contaminants that can affect quality of lakes and streams. More herbicides are used than any other pesticide, making it a high priority for assessing guidelines proposed by EPA and state regulatory agencies setting limits on what constitutes levels of pollutant that impairs surface water bodies (i.e., Total Maximum Daily Load, TMDL). A comprehensive evaluation of herbicide impacts on the environment needs to follow the fate of herbicides from the point of application in the field to entry and dissipation in surface waters.

Soil characteristics determine the quantity of herbicide to be applied and the rate at which that herbicide dissipates. Sorption of many herbicides to soil increases with clay and organic carbon (OC) content (1, 2, 3), thereby reducing herbicide efficacy in controlling weeds (4). However, when considering the environmental risk of herbicides, more restricted movement to non-target sites in soils with higher clay and OC content is beneficial. The spatial variability of atrazine sorption in soils from an Iowa field site was affected by landscape position (5), with the greatest sorption in depression areas. The magnitude of atrazine sorption correlated with OC, pH and to a lesser extent clay content.

The Mississippi River Delta is a major agricultural region devoted primarily to cotton, soybeans, corn, and rice. The Delta region extends along a narrow band on either side of the Mississippi River from Southern Missouri to the coast of Louisiana, and is comprised of approximately 8.5 million hectares. A current dilemma for farmers is to meet environmental concerns while producing a quality product with sustained profitability. Traditional row crop management in the Delta Region includes extensive tillage and pesticide use with little crop or weed residue covering the soil surface during fallow winter months. It is during this period that soils are most susceptible to erosion. In spite of high annual rainfall (114 to 155 cm), water shortages often occur at critical crop growth stages, and there is significant acreage in the Delta that is irrigated, further contributing to soil erosion loss. Increasing public concern and governmental regulation to minimize soil loss on erodible soils has prompted the development of production practices that conserve natural resources, preserve environmental quality, and sustain agricultural productivity.

An interagency cooperative research and demonstration project (Mississippi Delta MSEA) was established in 1994 with the purpose of assessing how agricultural activities influence Delta Region surface and ground water quality. The project also fits a national priority research objective for improving and restoring water resources. A second overall objective of the project was to increase knowledge about best management practices (BMP's) applicable to the Mississippi Delta region and quantify how they can improve soil and water resources. This project should bring the knowledge obtained from years of experimental plot studies to the farm level. Both environmental and economic benefits need to be quantified to demonstrate to producers how their farms can be improved by using BMP's.

In order to develop a comprehensive evaluation of herbicide dissipation in a Mississippi watershed, research focused on fluometuron [N,N-dimethyl-N'-[3-trifluoromethyl) phenyl urea]. The approach taken for this paper was to evaluate fluometuron dissipation in different settings ranging from the field where it was applied to a nearby water body. Fluometuron is the most widely used herbicide in cotton production in the southern United States. A recent survey of herbicides observed in three Mississippi Delta streams indicated that fluometuron was routinely detected from spring months when herbicides are often applied until early fall (6). Fluometuron is a member of the phenyl urea group of herbicides, and is used to control many broadleaf and grass weed species (7). It is applied to soil, usually before emergence of the cotton crop, but also can be applied in a directed spray on weeds after cotton emergence. The half-life of fluometuron in soil has been measured from 14 to 120 days after application (e.g., 8, 9, 10, 11).

Understanding the fate of metabolites as well as the parent compound is very germane to assessing environmental impacts of herbicides. Metabolite degradation is dependent on both the soil properties and chemical properties of the metabolite. The major metabolite of fluometuron in soil, desmethyl fluometuron (DMFM), results from N-demethylation (12). In the study by Coupe et al. (6), DMFM also was detected in the surface water. Further demethylation of DMFM forms trifluoromethyl phenyl urea (TFMPU), which is typically observed in low concentrations. Hydrolysis of either fluometuron or TFMPU will form trifluoro-methylaniline (TFMA), rarely reported in soil, but which has been detected in water samples in Texas (13).

Materials and Methods

Beasley Lake Watershed

The Mississippi Delta MSEA project is centered around three oxbow lake watersheds, i.e., old cut off river channel meanders (14). The approach of using oxbow lakes for the present project was chosen because it allows assessment of an entire, closed watershed. In some other MSEA projects, the watersheds were very large, making it difficult to evaluate how management practices were influencing bodies of water. In an oxbow lake watershed, all water drains into the lake, and lake health can be monitored over time.

The data presented in this paper are from only one of the watersheds, Beasley Lake. Beasley Lake watershed consists of 405 ha, and the lake is 17 ha in size. Beasley Lake is an oxbow lake that was once part of the Sunflower River (Figure 1). The topography of the region is relatively flat, with less than 1% slope toward the lake. A large, forested riparian area drains into the lake. The only improvements installed by researchers in Beasley Lake watershed were slotted board risers and grass filter strips. Slotted board risers are drainage outlets constructed so that during periods of heavy runoff, a board can be installed to slow water movement through the outlet, thus allowing sediment to settle. Vegetative buffer strips (primarily grass) along edges of fields were another physical improvement added at Beasley Lake.

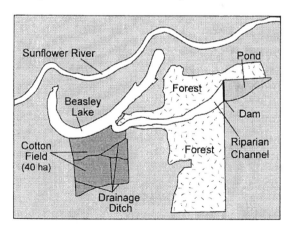

Figure1. Map of Beasley Lake watershed, including 40-ha cotton field where fluometuron dissipation was evaluated, forested riparian channel, and lake.

Study Site and Soil Sampling for Herbicide Dissipation

Cotton and soybean were the predominant crops grown in Beasley Lake from 1995 to 1997, the period of interest for this paper. Overall research objectives for the data presented here were to assess relationships among management practices, herbicide dissipation, soil resources, and lake health. For field dissipation studies, a 40-ha area under cotton production was selected (see outlined area in Figure 1). The area was managed with practices conventional for cotton production in the Mississippi Delta Region, including disking in the spring and fall, forming row beds in the spring, and cultivation after planting.

The study area contained a diverse number of soil types. Soil survey data indicated that Dundee (fine-silty, mixed, thermic Typic Endoaqualfs), Forestdale (fine, smectitic, thermic Typic Endoaqualfs), Dowling (very-fine, smectitic, thermic Vertic Epiaquepts) and Alligator (very-fine, smectitic, thermic Alic Dystraquerts) are major soil series represented. A 60 m by 60 m square grid was established, with column and row ends permanently marked and the grid locations georeferenced (Pathfinder ProXR, Trimble Navigation, Ltd., Sunnyvale, CA). Each grid node was the center of a 2 m x 2 m plot from which soil samples were collected.

At cotton planting in Spring, 1996, fluometuron (0.6 kg ha^{-1}) herbicide was applied to the soil in a band at the top of the row bed. Fourteen sampling sites were selected within the 40-ha area for evaluation of fluometuron dissipation, and each sampling site was centered on a grid node. The sample sites were chosen so that the soil types common to the watershed area were represented. Soil samples were collected from the 0-5, 5-15, and 15-25 cm depths in 1996 just prior to fluometuron application and periodically for four weeks after application. Previous knowledge of the approximate half-life of fluometuron gave an indication of the length of time the soil could be sampled to obtain a relative assessment of dissipation.

Soil Analysis

Soil samples for the fluometuron dissipation study were frozen until extraction and analysis. Field moist soil samples were extracted with methanol (2:1 soil:MeOH w:v). Extracts were centrifuged after a 24-h shaking, filtered with Gelman Acrodiscs (0.25 µm) (Gelman Laboratory, Ann Arbor, MI), and analyzed with HPLC. HPLC conditions included: HPLC System 2690 (Waters, Inc., Milford, MA); Fluorescence Detector 486 (Waters, Inc.); Econosil reverse-phase C18 column (Alltech Assoc., Deerfield, IL); gradient 55:45 water:ACN to

30:70; flow rate one mL min^{-1}; and fluorescence wavelengths Em 329, Ex 294. The chromatograms were assessed for fluometuron and a dominant metabolite, desmethyl fluometuron (DMFM). Standards of fluometuron technical grade and DMFM were obtained from Novartis, Inc., Greensboro, NC.

Fluometuron sorption was evaluated in soil samples from 50 of the grid node points (0-5 and 15-25 cm soil depths). The 0-5 cm soil depth was chosen because it is the depth of the soil profile that will likely have the most influence on herbicide dissipation. The 15-25 cm depth was selected as the interface between where mixing due to tillage occurred and the relatively undisturbed subsoil. Batch methods were used to assess fluometuron sorption. Fluometuron used for analysis included ^{14}C-labelled (98% purity) and technical grade (Novartis, Inc.). One fluometuron concentration (11.5 µM, 42 Bq mL^{-1}) in 0.01 M CaCl$_2$ was used, with three replications, and the experiment was repeated. Samples were equilibrated in 25-mL Pyrex centrifuge tubes (1:1.5 soil:solution, 5g:7.5 mL) for 17 h at 25 °C, centrifuged (10,000 x g, 15 min), and the supernatant filtered (Whatman 42 filter paper, Whatman, NJ). ^{14}C-labelled fluometuron remaining in solution after equilibration was counted for radioactivity (Tri-Carb 4000, Packard Instrument Co., Downers Grove, IL) after mixing with Eco-Lite scintillation cocktail (ICN, Costa Mesa, CA).

Agronomic Practices and Weed Counts

The Beasley site was conventionally tilled and irrigated. Weed management included initial burndown with glyphosate [N-(phosphonomethyl)glycine]. Following re-bedding of rows, fluometuron and metolachlor [2-chloro-N-(2-ethyl-6-methylphenyl)-N-(2-methoxy-1-methylethyl)acetamide] were applied pre-emergence (1.7 and 1.1 kg ha^{-1}, respectively) in a 43-cm band at planting. In 1996 and 1997, weed counts were taken six weeks following fluometuron and metolachlor application and prior to post-emergence application of cyanazine and MSMA. Weeds were counted in one-meter strips within 15 cm of the cotton row, since the fluometuron was applied in a band along the top of the row. The weeds were categorized as controlled or not controlled (7) by fluometuron and metolachlor.

Surface Movement of Herbicide

Fluometuron herbicide was applied in a band on top of the row beds (Figure 2a). A study was conducted to evaluate herbicide movement from the band at the top of the row bed, into the row furrow, and down the slope (Figure 2b). In

1996, three rows were selected that sloped (approximately 1% slope) 50 to 80 m toward a drainage ditch. The rows were spaced 30 m apart. From each row, four sampling sites were designated (Figure 2b): top of the slope, end of the row at the bottom of the slope, turn-row (2 to 3 m wide) area between the row end and a grassy area, and grassy area between turn-row and ditch (3 to 4 m wide). Soil was sampled from the surface 0 to 2 cm depth in the furrow between row beds (Figure 2a). Soil was assessed at that shallow depth because the primary interest was in surface movement of fluometuron. Fluometuron leached below 2 cm would not likely be moved in subsequent runoff events. Samplings were taken before fluometuron application, just after fluometuron application, 11 days after application, and 34 days after application. Rain and / or irrigation events occurred just prior to the last two samplings. Soil samples were frozen until extraction and analysis as described previously.

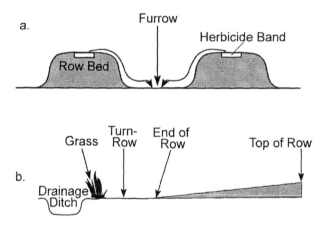

Figure2. Ilustrations of (a) row beds and herbicide band applied to the top of the row. Herbicide potentially could move from top of row bed into the furrow between row beds. (b) Soil samples were collected in the furrow, and sampling points include the top of the row, end of the row, turn-row area between the row end and grass, and grass strip.

Riparian Evaluations of Herbicide Degradation Potential

A large riparian forest adjoins Beasley Lake. This area is a wetland with a channel that runs through the middle, serving as a conduit for water draining from surrounding fields into the lake. Part of our objective was to assess the capability of the riparian area to filter contaminants moving in runoff toward the lake. Soil samples were collected from the field and three zones along the riparian channel toward Beasley Lake: 0 to 25 m from dam at edge of field, 50 to 200 m, and 400 to 800 m (Figure 1). Three enzyme assays were conducted to estimate heterotrophic microbial activity and the potential for herbicide degradation. Hydrolytic enzyme activity (esterase, lipase and protease) assays were conducted using fluorescein diacetate (FDA) as substrate (15). Dehydrogenase activity was assayed using triphenyl tetrazolium chloride (TTC) as substrate (3.0% aqueous with 0.1% yeast extract as an exogenous carbon source) (16). Aryl acylamidase was assessed using 2-nitroacetanilide (2-NAA) as substrate (17). All assays were conducted with three substrate replicates and one no-substrate control for each sample. Activities per hour were calculated based upon extinction coefficients and are reported as nmole of product formed (fluorescein for FDA, triphenyl formazan for TTC, and 2-nitroaniline for 2-NAA) g^{-1} soil (oven dry weight) h^{-1}. The potential for aerobic fluometuron degradation was determined in a soil slurry assay (1:10 dilution of soil, fresh soil weight). One mL of the soil dilution was transferred to sterile screw cap tubes and 1.0 mL of fluometuron solution (20 µg mL^{-1}) was added (five replicate tubes per sample). Tubes were incubated on a shaking incubator (25°C, 75 rpm) for 28 days. The study was terminated by adding 2 mL methanol, shaking at 75 rpm for an additional 24 h, centrifugation (10 min. 10,000 x g) and filtered (0.2 µm). Concentrations of fluometuron and the metabolite DMFM were determined by HPLC as described previously.

Evaluations of Herbicide in Lake Water

Surface (0 to 20 cm deep) water samples were collected monthly from three locations within Beasley Lake. Water samples (300 mL) were centrifuged and filtered through Empore C18 (Varian, Walnut Creek, CA) discs. The discs were eluted in 20 mL ethyl acetate and then concentrated to 2 mL. The concentrated samples were analyzed with HPLC as described previously.

Statistical Analyses

Geostatistics were used to assess the spatial variability of soil characteristics, weed populations, and fluometuron sorption (GS+, Gamma Design Software, Plainwell, MI). Linear or spherical models were used to describe these soil characteristics. These models, together with experimental data, were used to estimate average values (18) for soil properties and weed densities in square 0.36 ha areas centered about each sampling grid node.

Effects of soil properties on weed populations were examined with regressions and cross-semivariograms (18, 19). For further details on geostatistical analyses, readers are referred to Gaston et al. (4).

Results and Discussion

Soil Characterization and Fluometuron Sorption

Soil characteristics for the 40-ha field dissipation area are shown in Table I. In both surface and subsurface, there was a wide range in values, reflecting the alluvial origins of the soils in this watershed. Soil clay tended to be higher, and OC and sand lower in the 15 to 25 cm depth.

Table I. Soil characteristics in the surface and subsurface soil in the Beasley Lake watershed.

| Property | Soil Depth | | | |
| | 0 to 5 cm | | 15 to 25 cm | |
	Range	Mean (Std. Dev.)	Range	Mean (Std. Dev.)
% Clay	13 to 59	31.0 (10.7)	22 to 63	40.4 (9.9)
% Sand	1 to 49	17.5 (10.1)	1 to 39	11.2 (8.7)
% Organic C	0.51 to 2.53	1.67 (0.36)	0.07 to 0.96	0.47 (0.21)

Fluometuron sorption in the 0 to 5 cm depth ranged from 15 to 53 % of applied (mean 34.0 %; std. dev. 9.3), and in the 15 to 25 cm depth, sorption ranges were 18 to 58 % of applied (mean 35.0 %, std. dev. 9.1). Correlations were used to compare fluometuron sorption with other soil parameters, and for both 0 to 5 cm and 15 to 25 cm soil depths, the trends were the same (Table II). Organic C was positively correlated with organic matter. Sorption of substituted urea herbicides, such as fluometuron, in soil is believed to be due to a combination of mechanisms, primarily related to OC content. Nonionic mechanisms described as hydrophobic bonding (20, 21) or van der Waals attractions (22) likely play a role. Mechanisms such as hydrogen bonding (23, 24, 25, 26) may be involved with sorption to polar sites in soil.

In the Beasley watershed soils, a positive correlation between fluometuron sorption and clay content also was obtained, while the sand fraction was negatively correlated with fluometuron sorption (Table II). However, the role of clay in the sorption of substituted ureas is not clear, and poor correlations of urea sorption and clay content have been observed in other studies (20, 21, 23, 27, 28, 29, 30). One distinction between those studies cited and the present study is that these soils are from the same locale, Beasley watershed, and share common characteristics, such as mineralogy. It has been shown that the extent of complex formation between substituted ureas and clay minerals, such as montmorillonite, can depend on the characteristics of the cations associated with them (23, 31, 32), and it is likely that the montmorillonitic fraction in the Beasely soils was similar. Although some of the correlation may result from co-correlation between clay and OC, the correlation of fluometuron sorption with clay content increased in the 15 to 25 cm depth, while the correlation with OC decreased. The lower correlation in the 15 to 25 cm depth was attributed to a 70% lower OC content, and the higher correlation with clay was likely due to a higher clay content in the subsurface.

Table II. Correlation (r) between soil characteristics and fluometuron sorption in surface and subsurface soil in the Beasley Lake watershed.

Property	Soil Depth	
	0-5 cm	*15-25 cm*
	------------ $(r)*$ ------------	
Clay	0.65	0.72
Sand	-0.55	-0.55
Organic C	0.68	0.47

*Significant at 0.05 probability level.

Geostatistical models were used to describe semivariograms for the measured soil characteristics such as OC and texture (4) and for fluometuron sorption, which were then used to generate contour maps of the variables by kriging (Figure 3). The contour plots illustrate the variability of these alluvial soils. The lighter areas in Figure 3 represent lower values of a corresponding characteristic. Similarities in the contour patterns can be observed and support the correlations shown in Table II, especially for clay and herbicide sorption. Areas of higher clay and OC tended to occur in depressions or at the bottoms of slopes, reflecting depositions of sediment from runoff over a number of years. The patterns observed here also reflect the alluvial origins of these soils. The ridges, or lower clay areas, are essentially old sand bars formed when the river meandered back and forth across this area.

Spatial variability of soil characteristics can be used to assess biological parameters for a defined area. One agronomic parameter that Gaston et al. (4) studied in relationship to soil properties was weed population. Figure 4 illustrates a positive relationship between weed population and organic matter and clay. As either soil organic matter or clay contents increases, so does the number of potential exchange sites for any herbicide that is applied to the soil. This implies that in areas with high clay or organic matter, it would be necessary to apply more herbicide to attain weed control comparable to areas with lower clay or organic matter. The weed data shown in Figure 4 consist of total weeds. When only the weed species controlled by fluometuron was considered, spatial relationships were less clear. However, weeds tended to reoccur in areas where clay and OC were relatively high, indicating a relationship between herbicide persistence and soil properties. However, relatively coarse textured areas were commonly weed-free for two years. These observations suggest that greater uniformity of weed control might be achieved by a variable rate of herbicide application. Also, adequate weed control might be achieved at a reduced rate of application in sandy, low OC areas.

Fluometuron Field Dissipation

Field dissipation of fluometuron was measured to evaluate relationships between soil characteristics and herbicide dissipation. Correlations between the half-life of fluometuron in the surface soil and several soil parameters are presented in Table III. There was a significant positive relationship between fluometuron dissipation and clay content, with a negative relationship for the coarser textured soil fractions. As a result of the significant correlations of herbicide dissipation with clay and sand, soils were categorized into two groups based on clay content. The values for each group were averaged over the course of the 28-day assessment period and are shown in Figure 5. More rapid herbicide dissipation was observed in soils with lower clay content, and this was attributed to several factors. There is the potential for some sequestration or protection from degradation when herbicide is sorbed (11). Coarser textured soils are more aerated, providing an environment conducive for oxidative

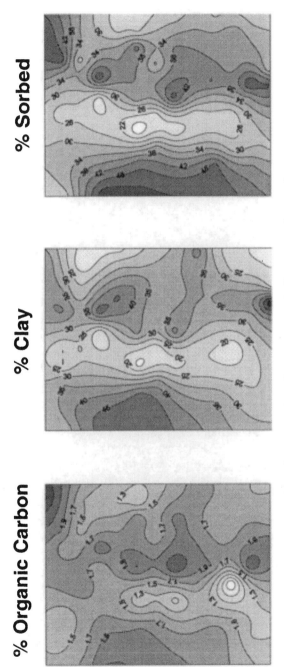

Figure 3. Soil contour plot of percent organic C, percent clay, and percent fluometuron sorption in a cotton production area, Beasley Lake watershed. The soil depth is 0 to 5 cm.

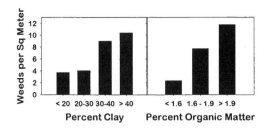

Figure 4. Effect of soil clay and organic matter content on control of weeds (adapted from Gaston et al., 2001).

processes such as *N*-dealkylation of fluometuron. Another consideration is that there is less fluometuron retention in the coarser textured soils, and the herbicide is more subject to movement away from the site of application by leaching or surface runoff.

No relationship was observed between fluometuron dissipation and OC (Table III). Several conflicting factors likely contributed to a lack of correlation between herbicide dissipation and OC content. Under optimum moisture conditions, higher OC levels in surface soils are often associated with enhanced microbial activity. High microbial activities can lead to co-metabolism of xenobiotics in soil. The potential, therefore, exists for a positive relationship between OC levels and herbicide dissipation. However, surface soils often are dry crusts, even during the traditionally wet Spring when the fluometuron was applied. The dry soil condition would severely reduce microbial activity and accompanying herbicide metabolism. Also, in soils with higher OC, there may have been some protection from degradation or lowered mobility because of enhanced herbicide sorption. Another consideration is that, while there was a wide range in OC levels for the surface soils in this watershed, the average OC level (Table I), is relatively low and may not be sufficient to enhance microbial activity to any significant degree.

Table III. Correlation (r) of surface soil properties with fluometuron dissipation (half-life) and average DMFM ($\mu g\ g^{-1}$) concentration.

Soil Property	Fluometuron (half-life)		DMFM (average conc.)	
	r	Prob. > [r]	R	Prob. > [r]
Clay	0.62	0.05	0.52	0.06
Sand	-0.46	0.10	-0.29	NS
Silt	-0.57	0.05	-0.61	0.05
OC	0.28	NS	0.23	NS

Note: NS = Not Significant

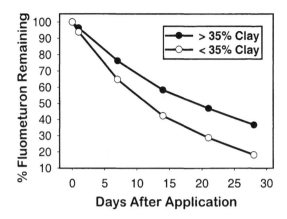

Figure 5. Effect of clay content on fluometuron dissipation from surface (0-5 cm) soil.

In the 5 to 15 cm soil depth, fluometuron was detected erratically in some plots, usually later than two weeks after application, and for only a short period of time (data not shown). No trends related to soil characteristics were observed in the 5 to 15 cm samples. Herbicide was not detected at depths below 15 cm. Herbicide sorption is often greater in surface than sub-surface soil (1, 10, 33, 34). In most of these studies, the increased sorption capacity of surface soils was primarily attributed to a greater mass of organic carbon (greater number of sorption sites) rather than to a stronger affinity or energy of sorption by organic components in the soils. Enhanced fluometuron sorption in the surface would reduce mobility to greater soil depths, and this may have been a factor in the low levels of fluometuron measured in soil at depths below 5 cm. The absence of fluometuron in soil , however, does not conclusively demonstrate that fluometuron did not leach. For example, Essington et al. (35) measured fluometuron in leachate collected at 90 cm soil depths, attributing movement primarily to preferential flow.

Demethylated fluometuron or DMFM is one of the primary metabolites of fluometuron measured in soil. For each sampling area, the DMFM concentration over the 28-day period was averaged and correlated with various soil characteristics (Table III). Similar to fluometuron, there was a positive correlation for clay and a negative relationship for silt, but no relationship was significant for sand or organic C.

The soils were grouped according to clay content and the average DMFM concentrations in soil (Figure 6). The pattern for DMFM in surface soil indicates that it does not accumulate, but the peaks coincide with the decline in fluometuron concentration, and then the DMFM dissipates. The soils with higher clay generally accumulated more DMFM than lower clay soils. While

DMFM sorption was not evaluated in this watershed, in another study (36), DMFM sorption was measured for soils of varying soil properties. Dundee silt loam and Tunica clay had similar organic carbon levels (1.7 and 1.5 %, respectively), but different clay contents (21 and 55 %, respectively). The Tunica clay is predominantly montmorillonitic and shows higher DMFM sorption ($K_{Freundlich}$ 3.7, N=0.76) than the Dundee silt loam ($K_{Freundlich}$ 3.0, N=0.77). As with fluometuron, some of the same arguments could be made that greater sorption in the higher clay soils reduced the mobility and enhanced protection from degradation.

Surface Movement of Fluometuron

Fluometuron concentrations in the soil surface (0 to 2 cm) at various sampling locations are presented in Figure 7 during the first month after application. Baseline (before herbicide application) fluometuron concentrations were negligible in all four sampling locations, providing an excellent reference point from which to evaluate subsequent changes. Fluometuron concentration was detected in all locations two days after application even though no measurable precipitation was recorded. This may have been due to overspray or drift. After the first rain and an irrigation with a total of 3 cm water, fluometuron at the top of the slope and end of slope increased. Precipitation fell (11 cm) prior to the last sampling. The most dramatic increase in fluometuron concentration was measured in the turn-row area. The turn-row was a convergence area for all runoff flowing down the row slope, not just for the rows that were evaluated, but for all rows. As such, the turn-row area was a mixing zone and the fluometuron measured likely was from several sources. During heavy rain, the water ponded in the turn-row until it either could seep through or flood over the grass area to the ditch. Soil depths greater than 2 cm were not evaluated in the turn-row to determine the extent to which fluometuron may have leached.

It is noteworthy that relatively little fluometuron was measured in the surface soil from the grass area, indicating the effectiveness of the grass strip as a physical barrier restricting sediment or water movement. Also, fluometuron may have moved by mass flow to the ditch during times of flooding, with little residence time in the grass area. Additionally, there were breaks in the grass strip and eroded areas that would have provided channels for excess water to drain from the turn-row area to the ditch while bypassing the grass area. Some infiltration of runoff may have occurred in the grass area, which could have lowered outflow concentrations of herbicide due to sorption to soil or vegetative material (37, 38). Also, enhanced degradation of herbicide can occur in grass areas if residence time is sufficient (39, 40).

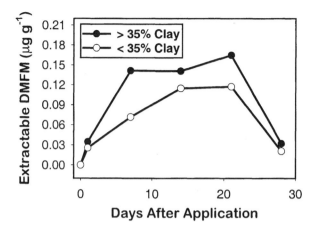

Figure 6. Effect of clay content on demethylated fluometuron (DMFM) dissipation from surface (0-5 cm) soil.

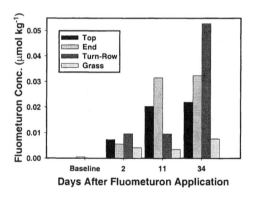

Figure 7. Surface movement of fluometuron in soil (0 to 2 cm) within a row furrow.

Riparian Area and Lake

Three enzyme activities, esterase, dehydrogenase and aryl acylamidase, were significantly greater in riparian zone soil compared to cropland soil (Table IV). Esterase, dehydrogenase, and aryl acylamidase activities were about 8 to 18, 3 to 12, and 1.5 to 3 fold-greater in riparian soil compared to the cultivated soil, respectively. The highest levels of all three enzymes were observed in the zone closest to the lake that had the highest soil moisture, OC, and clay contents. Cooper et al. (41) found similar conditions where the textural composition of soil near the entry point into forested riparian areas was sandier, while soil further into the riparian zone was predominantly silt and clay. Soil enzymatic activity can be used either in a lower resolution application as an index of general soil microbial activity or may be used for a higher resolution application for understanding a specific process (42). FDA hydrolytic activity represents a wide range of hydrolytic activity (esterase, lipase and protease) activity and correlates with soil respiratory activity. Dehydrogenase activity actually measures electron transport system activity in that the substrate TTC is used as an alternative electron acceptor (43), under both aerobic and anaerobic conditions. Both FDA-hydrolysis and TTC-dehydrogenase thus described the generic spatial variation in microbial activity due to position along the riparian zone. Aryl acylamidase activity assesses hydrolytic cleavage of an amide bond and thus can be used to describe a process such as herbicide degradation, e.g., acylanilide and phenylurea herbicides.

Table IV. Soil enzyme activity and *in* vitro fluometuron degradation in soils collected along a transect of the Beasley Lake riparian zone.

Sampling Location	FDA-esterase	TTC dehydro-genase	2-NAA aryl acylamidase	DMFM Formed
	nmol formed g^{-1} $soil^{-1}$ h^{-1}			nmol g^{-1}
Cultivated Field	133 ± 44*	1.8 ± 0.8	28 ± 4	28 ± 5
Riparian Zone 1 (Dam to 25 m)	1507 ± 566	6.1 ± 0.8	42 ± 6	57 ± 17
Riparian Zone 2 (50 to 200 m)	1106 ± 93	6.4 ± 3.3	67 ± 22	62 ± 13
Riparian Zone 3 (400 to 800 m)	2361 ± 494	22.7 ± 14.7	88 ± 33	111 ± 27

*Mean ± standard deviation of three replicates

The potential for soils to degrade fluometuron in a soil slurry system was studied by monitoring both dissipation of the parent compound and accumulation of the metabolites. The DMFM metabolite accumulated in all soils with the greatest accumulation in soil suspensions from the riparian zones (Table IV). Patterns of fluometuron dissipation exhibited a similar trend as DMFM accumulation (data not shown). Studies by Entry and Emminham (44) indicated that atrazine and 2,4-D were more rapidly degraded in riparian zones under coniferous forests compared to grasslands. Higher microbial activity and populations observed in Mississippi Delta riparian zones should facilitate rapid degradation of pesticides. Preservation and maintenance of forested riparian zones should be an important management practice to reduce non-point pollution of surface waters by agrochemicals.

There is concern that bodies of water, such as lakes, are the sink for agrochemicals such as pesticides and nutrients. Questions are raised as to the extent these chemicals are reaching the lakes and whether management practices are impeding movement of chemicals into the lakes. Table V shows the concentrations of fluometuron and DMFM in lake water during the period from May to October. This period coincides with the time just after fluometuron application and subsequent dissipation in the field. Allowing for a lag time in May, during which the fluometuron moved to the lake in sufficient quantity for detection, the occurrence of fluometuron in the lake mirrored what was observed in the field. There is a gradual buildup in fluometuron concentration in the lake during June through July, with a peak in July. Concentrations then decline to undetectable levels from September to October. DMFM followed a similar pattern, but at lower concentrations (Table V). No TFMA was observed in any samples from these studies, in contrast to other reports (6, 13). Concentrations and dynamics of appearance of fluometuron and DMFM are similar to those observed in Mississippi Delta streams (6). Vegetative filter strips and a large riparian zone may have entrapped and impeded fluometuron movement, thus delaying fluometuron appearance until over a month after application. DMFM generated in the field during fluometuron degradation is subject to movement from the field to the lake. DMFM is likely less mobile than fluometuron, based on its higher sorption K value relative to fluometuron (45). It is likely, therefore, that much of the DMFM measured in lake water is a result of *in situ* metabolism of fluometuron, e.g., algal *N*-dealkylation (46).

Table V. Fluometuron concentrations in Beasley Lake, 1997

Month	Fluometuron	DMFM
	$\mu g \, L^{-1}$	$\mu g \, L^{-1}$
May	<0.1	<0.1
June	3.2 ± 1.9	0.8 ± 0.6
July	5.7 ± 3.4	2.0 ± 0.5
August	4.8 ± 2.2	1.8 ± 0.6
September	0.7 ± 0.6	0.7 ± 0.6
October	<0.1	<0.1

References

1. Locke, M.A. *J. Environ. Qual.* **1992,** *21,* 558-566.
2. Locke, M.A.; Gaston, L.A.; Zablotowicz, R.M. *J. Agric. Food Chem.* **1996,** *44,* 1128-1134.
3. Locke, M.A.; Gaston, L.A.; Zablotowicz, R.M. *J. Agric. Food Chem.* **1997,** *45,* 286-293.
4. Gaston, L.A., Locke, M.A.; Zablotowicz, R.M.; Reddy, K.N. *Soil Sci. Soc. Am. J.* **2001,** *65,* In press.
5. Novak, J.M.; Moorman, T.B.; Cambardella, C.A. *J. Environ. Qual.* **1997,** *26,* 1271-1277.
6. Coupe, R.H.; Thurman, E.M.; Zimmerman, L.R. *Environ. Sci. Technol.* **1998,** *32,* 3673-3680
7. Mississippi State University Extension Service. Weed control guidelines for Mississippi. Mississippi State University, Mississippi State, MS, 2001; Pub. 1532, 197 pp.
8. LaFleur, K.S.; Wojeck, G.A.; McCaskill, W.R. *J. Environ. Qual.* **1973,** *2,* 515-518.
9. Rogers, C.B.; Talbert, R.E.; Mattice, J.D.; Lavy, T.L.; Frans, R.E. *Weed Sci.* **1985,** *34,* 122-130.
10. Brown, B.A.; Hayes, R.M.; Tyler, D.D.; Mueller, T.C. *Weed Sci.* **1994,** *44,* 171-175.
11. Zablotowicz, R.M.; Locke, M.A.; Gaston, L.A.; Bryson, C.T. *Soil and Tillage Res.* **2000,** *57,* 61-68.
12. Bozarth, G.A.; Funderburk, H.H. *Weed Sci.* **1971,** *19,* 691-695.
13. Thurman, E.M.; Bastian, K.C.; Mollhagen, T. *Sci. Tot. Environ.* **2000,** *248,* 189-200.
14. Cooper, C.M.; Rebich, R.A.; Locke, M.A. In *Proc. 3rd Natl. Workshop on Constructed Wetlands Conference / BMPs for Nutrient Reduction and Coastal Water Protection,* USDA-CSREES, EPA, USDA-ARS, CTIC, and ASAE, 1999, pp. 33-34.
15. Schnürer, J.; Rosswall, T. *Appl. Environ. Microbiol.* **1982,** *43,* 1256-1261.
16. Casida, L.E., Jr. *Appl. Environ. Microbiol.* **1977,** 34, 630-636.
17. Zablotowicz, R.M.; Hoagland, R.E.; Wagner, S.C. *Soil Biol. Biochem.* **1998,** *30,* 679-686.
18. David, M. *Geostatistical ore reserve estimation*; Elsevier Scientific Publ. Co., New York, NY, 1977.
19. Yeats, S.R.; Warrick, A.W. *Soil Sci. Soc. Am. J.* **1987,** *51,* 23-30.
20. Carringer, R.D.; Weber, J.B.; Monaco, T.J. *J. Agric. Food Chem.* **1975,** *23,* 568-572.
21. Kozak, J.; Weber, J.B. *Weed Sci.* **1983,** *31,* 368-372.
22. Hance, R.J. *Weed Res.* **1969,** *9,* 108-113.

23. Bailey, G.W.; White, J.L.; Rothberg, T. *Soil Sci. Soc. Am. Proc.* **1968**, *32*, 222-234.
24. LaFleur, K.S. *Soil Sci.* **1973**, *116*, 376-382.
25. Bouchard, D.C.; Lavy, T.L.; Marx, D.B. *Weed Sci.* **1982**, *30*, 629-632.
26. Bouchard, D.C.; Wood, A.L. *Toxicol. Indust. Health* **1988**, *4*, 341-349.
27. Hance, R.J. *Weed Res.* **1965**, *5*, 108-114.
28. Savage, K.E.; Wauchope, R.D. *Weed Sci.* **1974**, *22*, 106-110.
29. Grover, R. *Can. J. Soil Sci.* **1975**, *55*, 127-135.
30. Peck, D.E.; Corwin, D.L.; Farmer, W.J. *J. Environ. Qual.* **1980**, *9*, 101-106.
31. Hance, R.J. *Weed Res.* **1971**, *11*, 106-110.
32. Khan, S.U. *Can. J. Soil Sci.* **1974**, *54*, 235-237.
33. Locke, M.A.; Zablotowicz, R.M.; Gaston, L.A. In *Conservation farming: A focus on water quality*; Kingery, W.L; Buehring, N. Eds.; MAFES Special Bull. 88-7, 1995, pp. 55-58.
34. Reddy, K.N.; Locke, M.A.; Wagner, S.C.; Zablotowicz, R.M.; Gaston, L.A.; Smeda, R.J. *J. Agric. Food Chem.* **1995**, *43*, 2752-2757.
35. Essington, M.E.; Tyler, D.D.; Wilson, G.V. *Soil Sci.* **1995**, *160*, 405-414.
36. Locke, M.A.; Zablotowicz, R.M. *Weed Sci. Soc. Am. Abstr.* **2000, *40*,** 88.
37. Arora, K.; Mickelson, S.K.; Baker, J.L.; Tierney, D.P.; Peters, C.J. *Trans. Am. Soc. Agric. Eng.* **1996**, *39*, 2155-2162.
38. Misra, A.K.; Baker, J.L.; Mickelson, S.K.; Shang, H. *Trans. Am. Soc. Agric. Eng.* **1996**, *39*, 2105-2111.
39. Benoit, P.; Barriuso, E.; Vidon, P.; Real, B. *J. Environ. Qual.* **1999**, *28*, 121-129.
40. Staddon, W.J.; Locke, M.A.; Zablotowicz, R.M. *Am. Soc. Agron. Abstr.* **1998**, *90*, 31.
41. Cooper, J.R.; Gilliam, J.W.; Daniels, R.B.; Robarge, W.P. *J. Environ. Qual.* **1987**, *51*, 416-420.
42. Sinsabaugh, R.L. *Biol. Fert. Soils.* **1994**, *17*, 69-74.
43. Trevors, J.T. *CRC Critical Rev. Microbiol.* **1983**, *11*, 83-100.
44. Entry, J.A.; Emmingham, W.H. *Can. J. Soil Sci.* **1996**, *76*, 101-106.
45. Zablotowicz, R.M.; Locke, M.A.; Knight, S.S. In *Proc. Miss. Water Resources Researcg Conf.*; Ballweber, J.A. Ed.; Miss. Water Resources Res. Instit., 2000, pp. 38-42.
46. Locke, M.A.; Zablotowicz, R.M. *Am. Soc. Agron. Abstr.* **2000, *40*,** 88.

Chapter 15

Leaching of Post-Emergence Herbicides into Shallow Groundwater during Transition from Conventional Tillage to No-Tillage on Des Moines Lobe Soils of Central Iowa

T. R. Steinheimer and K. D. Scoggin

ARS National Soil Tilth Laboratory, Agricultural Research Service, U.S. Department of Agriculture, 2150 Pammel Drive, Ames, IA 50011

In a four-year field study carried out on a 96-ha field, we have evaluated the impact on shallow groundwater quality of transition from conventional-till to no-till for two herbicides now commonly used in minimum-till corn/soybean systems. Nicosulfuron and imazethapyr were studied. Well water samples were obtained from shallow piezometers using manual bailers and residue analyses carried out in our laboratory. Methods utilized SPE cartridge chemistry followed by HPLC/MS/MS techniques, providing a 10 part per-trillion quantitation limit in two-liters of filtered groundwater. By January of 1993, ten months following initial application, imazethapyr, nicosulfuron, and one nicosulfuron degradate had been detected in the well water and confirmed by MS/MS. During the 1993 through 1995 sampling seasons, average concentration in the piezometer samples was 131 and 125 part-per-trillion for nicosulfuron and imazethapyr, respectively. The data show that these widely used ALS-inhibiting chemicals are capable of leaching to shallow groundwater during tillage transition.

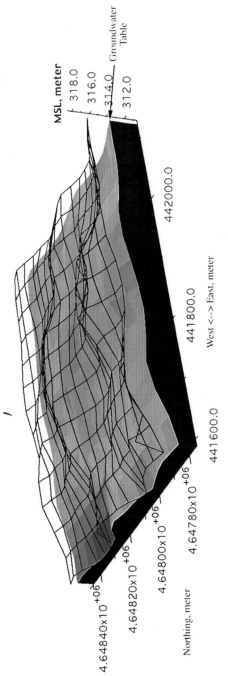

Plate 1. (a) Landscape Topography and Water-Table Depths for the 96-ha Field in WCW During Transition to No-Till in 1993; (b) Piezometer Locations.

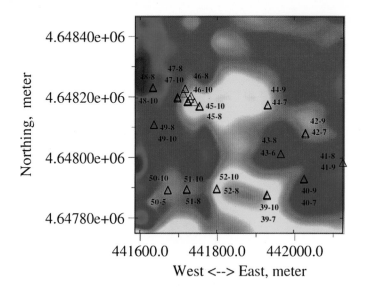

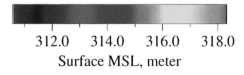

Surface MSL, meter

Plate 1. *Continued.*

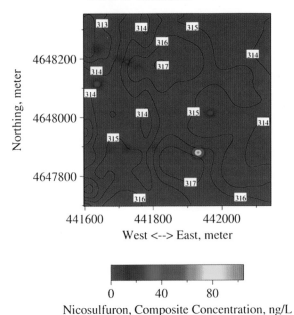

Nicosulfuron, Composite Concentration, ng/L

Plate 3. Composited Frequency of Detects Across the Field for (a) Nicosulfuron; (b) J290; (c) Imazethapyr.

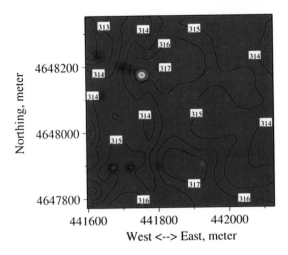

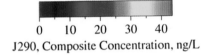

J290, Composite Concentration, ng/L

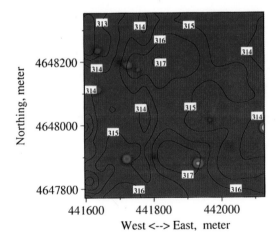

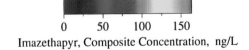

Imazethapyr, Composite Concentration, ng/L

Plate 3. *Continued.*

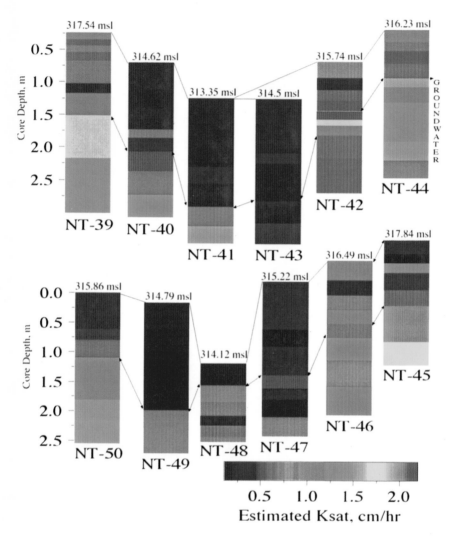

Plate 2. Estimated Variation in Saturated Hydraulic Conductivity with Depth Using 1994 Soil Cores.

Introduction

Agribusiness is the largest sector in the U.S. economy accounting for more than 30% of the Gross National Product. Major factors responsible for such growth have included the successful application of chemical-based fertilization and pest-control strategies to farming. In 1991, Congress funded the Presidential Initiative to Enhance Water Quality that was implemented nationwide. Two central objectives were identified: (1) measure the impact of farming systems on ground water and surface water chemistry as well as other agroecosystem resources, and (2) identify the factors and processes controlling the fate and movement of fertilizers and pesticides. In Iowa and four other midwestern states, field research was conducted under the Management System Evaluation Area (MSEA) program, a federal interagency, state, academia, cooperative study of best management practices (BMP's) and water quality (1,2). In Iowa, BMP's are defined in the context of combinations of tillage, crop rotation, crop sequencing, fertilizer application, and pesticide usage; whereas, in NE and MN, BMP's also include irrigation. Within the Iowa program, three areas under study represent diverse scales, landscapes, soil associations, and management practices. Two study areas which are operated by Iowa State University are located at the Northeast Research Station near Nashua and the Till Hydrology Site near Boone. A third, the Deep Loess Research Station near Treynor, is owned and operated by the Committee for Agricultural Development (CAD) and managed under a landlord/tennant relationship with two cooperating farmers. The MSEA program also required that a watershed-scale site be selected, and following initial survey and reconnaissance of the Des Moines Landform region, Walnut Creek Watershed (WCW) was chosen. By 1991, implementation of a monitoring effort was underway. In each, fields under controlled agronomic practices were extensively instrumented for monitoring environmental conditions. Throughout the year, both soil and water samples were collected for laboratory determination of chemical residues as the water quality legacy of both pesticide and fertilizer usage.

Iowa MSEA at the Watershed-Scale

The impetus for selecting a watershed-scale study was to provide an evaluation of the on-site and off-site movement of herbicides and nitrate-nitrogen from fields with different farming practices (3). Land-use in WCW is primarily row crop agriculture. The watershed is extensively tile drained with both surface and subsurface water draining directly into Walnut Creek. Open stream channels exist only in the lower half of the watershed. For MSEA studies, the watershed was divided into three subbasins, and within each, fields were selected for evaluation of different farming practices. Subbasins also provided a means for monitoring the integrative impact of farming practices. Farmers in the watershed were very cooperative in permitting MSEA project researchers to monitor their operations even though they often conflicted with and required rearrangement of scheduled farming activities.

Soils within Walnut Creek Watershed (WCW)

Soils within the WCW are formed predominantly in calcareous glacial till deposited within the Des Moines Lobe during the Cary substage of the Wisconsin glaciation (4,5). The till of the Des Moines Lobe is characterized by low relief swell and swale topography that is oriented transverse and slightly concave to the direction of ice flow. Within Walnut Creek, the swell swale orientation is southwest to northeast with an average relief of several meters. Surface drainage within the Des Moines Lobe is poorly developed due to the low relief and the area's geologic youth. Numerous closed depressional areas, commonly called "prairie potholes", exist and hold accumulated material from surrounding side slopes. The thickness of the accumulated material can range up to several meters. The western half of WCW is typical of this low relief till plane. However, the eastern portion of the watershed is characterized by increased surface relief of up to 10 meters as a result of head cutting of streams leading to the lower Skunk River floodplain. Soils within WCW are characterized by the Clarion-Nicollet-Canisteo association. This association consists of well drained [0.9-1.8 mm / hr, Ksat] Clarion and Lester soils located on higher or sloping areas, somewhat poorly drained [0.4 mm / hr, Ksat] Nicollet soils located on convex sideslopes, poorly drained [0.5-0.1 mm / hr Ksat] Canisteo and Webster soils located on low areas and drainage ways, and very poorly drained [0.04-0.6 mm / hr, Ksat] Okoboji and Harps soils located in closed depressional areas.

Surface and Subsurface Drainage

Drainage ditches, tile lines, and surface inlets connected to tile lines all have been added throughout WCW in order to increase its naturally poor drainage [long residence times] (6). Within the western half of the watershed the closed depressional areas are frequently filled with ponded water during periods of high precipitation. These pothole areas, in order to be farmed, have been systematically drained by these devices during the past 100 years. Fields closest to natural drainage ways were the first to be tile drained. Legislation in 1904 enabled the establishment of drainage districts which accelerated larger scale drainage. Between 1910 and 1920, still larger county tiles were installed as the framework for draining the depressions in fields located farther removed from the drainage channels. Field and county tiles installed from the late 1800's through the 1950's were made of clay and installed in approximately 0.60 meter sections that were placed end-to-end in trenches with a small gap between each. Tiles installed since are fabricated from perforated plastic tubing. Today, field tile diameters range from 12.7 to 40.6 cm depending on the drainage area. Most field tiles drain into county tile lines; however, some discharge directly into a drainage ditch or into Walnut Creek itself. Main county tile diameters range from 25 to 91 cm and increase in diameter along their drainage route in order to accommodate additional discharges from in-field county lines. In very poorly drained areas of the watershed, field intakes are installed to direct surface water to subsurface tile drains. In addition to subsurface drainage, elevated straight roads have been constructed and deep ditches dug on either side to direct water to intakes installed in the ditches. These county roads have affected the natural surface runoff direction by dividing many of the low areas into several sections and also have, in effect, created "dams" along the original swales and drainageways.

Cropping Patterns

Approximately 95% of the 5130 ha in WCW are used for row crop agriculture with limited livestock production. Land not used for row crop agriculture is pasture, woodland, or residential. The upper three-quarters of the watershed is relatively level and almost entirely in row crop production. Row crops, pasture, and woodlands are more evenly distributed in the lower quarter of the watershed, which has a more rolling topography and is more susceptible to erosion. Corn and soybeans are grown on 80% of the row crop acres in yearly rotations, although some continuous corn is also found in the watershed. More than 50% of the crop acres are commonly planted to corn and 40-45% are planted to soybeans (3,7). Alfalfa, grass and oats are grown on the remaining crop acreage that is usually involved with a Government land management program. Conventional tillage practices are used most commonly by farmers in the watershed. Management practices that include no-tillage have recently been adopted by some farmers, but neither this practice nor ridge tillage have been widely adopted. Chisel plows are used by more than 90% of the farmers in the watershed (1). Moldboard plows, however, are used by fewer than 5% of the farmers. A typical conventional farming operation for a corn/soybean field in WCW is summarized in Table I.

Transition to No-Tillage

Adoption of no-tillage practices has been promoted as a more sustainable farming practice because tillage often decreases or at least prevents the accumulation of soil organic matter (carbon). The increased plant residue remaining on the soil surface with no-tillage can result in either increased or at least maintenance of soil carbon levels, thus helping to sustain or enhance the soil quality. In addition, increased crop residue cover decreases the wind erosion potential, may increase soil microfauna and macrofauna, usually increases infiltration by promoting aggregate formation, reduces water evaporation at the

Table I. Sequence of Farming Operations for Typical Conventional Tillage Practices under Corn / Soybean Rotation in Central Iowa.

	Managment
Fall	chisel plow following corn harvest
Spring	seed bed preparation for soybean; field cultivation (cultivator), disking
Spring	soybean planting; 30 (or 15) inch rows; 180,000 seeds per acre
Spring	post-emergence application of herbicide
Early Summer	cultivation and / or rotary hoeing for weed control and breakup of crusted soil
Fall	Soybean harvest with combine
Late Fall	no tillage operations on soybean residue; overwinter
Late Fall - Early Spring	nitrogen fertilizer application knifed-into corn residue
Spring	seed bed preparation for corn; field cultivation (cultivator), disking
Spring	corn planting; 30 inch rows; 27,000 - 32,000 seeds per acre
Spring	post-emergence application of herbicide
Early Summer	cultivation of corn to reduce weed pressure and / or herbicide application; second nitrogen fertilizer application, if needed, based on LSNT (late spring nitrate test)
Fall	corn harvest with combine

soil surface, and minimizes the erosion runoff potential. Each of these factors carries a benefit that contributes to environmentally benign and sustainable farming (8,9). Furthermore, farmers benefit economically by reducing both their fuel costs and time necessary for cultivation. Among some farmers, however, there is concern that no-tillage will result in lower crop yields.

To evaluate the effect of transition to no-tillage on Des Moines Lobe soils, a 96 ha field located within northwestern WCW was chosen for this experiment. Landscape topography, piezometer locations, and water table depths within the field are detailed in Plates 1a and b. Prior to 1990, the field had been in long term conventional chisel-plow tillage. Starting with Spring, 1992, the field was put into no-till management. For planting, the field was divided approximately in half and seeded in both corn and soybean. The no-till farming operation is summarized in Table II. Rotating the crops to the other half the following year permitted the use of both one grass herbicide for corn and one grass herbicide for soybean on the same field simultaneously each year. This split-rotation of corn and soybean continued from 1992 through 1995. The objective of this study was to determine the potential for movement of herbicides and nitrate into surface and shallow ground water under no-till conditions. By the early 1990's a new generation of herbicides was gaining widespread use as weed-control tools for both corn and soybeans. The new chemistries promised season-long control with high-efficacy at low application rates and with minimum impact on other environmental resources (10,11). Because of their high activity they were also seen as fully compatible with conservation tillage practices, including no-till. These include both the sulfonylurea and the imidazolinone families.

Table II. Sequence of Farming Operations for Typical No-Tillage Practices under Corn / Soybean Rotation in Central Iowa

	Managment
Fall	corn harvest with combine
Spring	burn-down application of broad spectrum short term defoilant and application of pre-emergence soybean herbicide
Spring	soybean drilling in 8 - 10 inch rows; 180,000 seeds per acre
Spring	post-emergence application of herbicide
Fall	Soybean harvest with combine
Late Fall	no tillage operations on soybean residue; overwinter
Late Fall - Early Spring	nitrogen fertilizer application knifed-into corn residue
Spring	burn down application of broad spectrum short term defoilant and application of pre-emergent corn herbicide
Spring	corn planting, 30 inch rows; 31,000 - 35,000 seeds per acre
Spring	post-emergence application of herbicide
Early Summer	second application of nitrogen, if necessary, based on LSNT (late spring nitrate test)
Fall	corn harvest with combine

Herbicide Usage

By the mid-1990's, Accent[R] and Pursuit[R] were two of the fastest growing new post emergence herbicide formulations primarily for grass control in the central midwestern cornbelt of Iowa, the Dakota's, central Nebraska, the western half of Illinois, the southern half of Minnesota, and some areas of southern Wisconsin. Both were marketed as being fully compatible with conservation tillage practices being promoted at the time. Nicosulfuron, the active compound in Accent[R], is intended for use on field corn; imazethapyr, the active compound in Pursuit[R], is intended for use on soybean and on a few

SO2NHCONH = " SU "

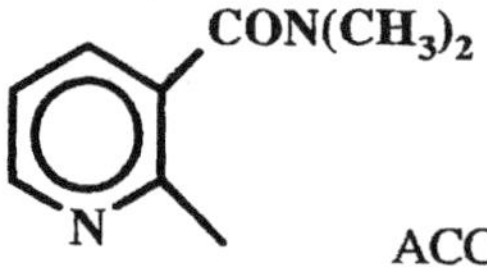

CON(CH$_3$)$_2$

ACCENT [nicosulfuron]

OCH$_3$

OCH$_3$

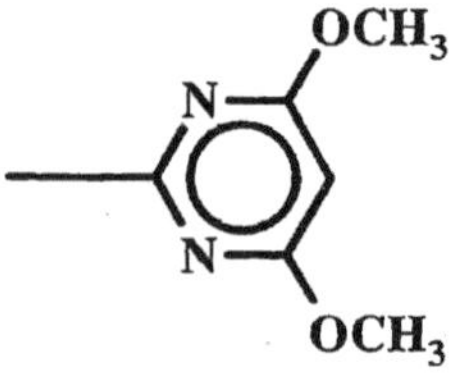

pH Dependent Water Solubility

Week Acid with pKa of 4.30

Soil Sorption Range : 0.1 to 2.5

Field Dissipation : 3 to 8 Weeks

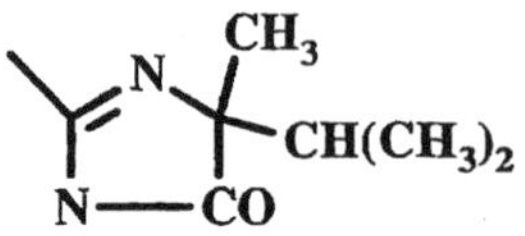

CH$_3$

CH(CH$_3$)$_2$ = " IMI "

N—CO

$_5$H$_2$C

CO$_2$H

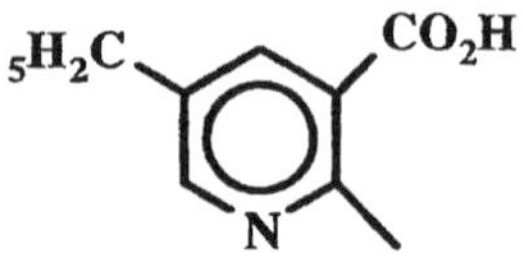

PURSUIT [imazethapyr]

1,400 PPM Water Solubility

Week Acid with pKa of 3.90

Soil Sorption Range : 0.4 to 0.8

Soil Metabolism : 33 to 37 Months

Field Dissipation : 17 to 340 Days

Figure 1. Chemical Structure and Environmental Fate Properties for Nicosulfuron and Imazethapyr.

imidazolinone-tolerant field corn hybrids. Figure 1 shows the chemical structure and dominant properties affecting environmental fate for each chemical. Nicosulfuron usage on corn increased from 11,300 kg in 1992 to 24,500 kg in 1995 in Iowa, and, in Illinois, from 3,600 kg in 1992 to 17,200 kg in 1995 (12-15). Similarly, imazethapyr usage on soybean increased from 103,600 kg in 1992 to 178,600 kg in 1995 in Iowa, and, in Illinois, from 67,200 kg in 1992 to 155,400 kg in 1995 (12-15). In addition, imazethapyr usage on imidazolinone-tolerant corn hybrids has grown in recent years. By 1994, Accent[R] usage in Iowa had grown to 19,500 kg and Pursuit[R] usage had grown to 154,500 kg (14,15). Given the recommended application rates at the time these kg quantities translated into approximately 2.5 - 5.0 million and 1.0 - 2.5 million ha for corn and soybean, in Iowa, respectively. Beginning with the Spring planting of 1992, metolachlor was applied as an in-row band followed some 2-3 weeks later by a post-emergence broadcast application at the maximum label-rate for Accent[R] over the corn and for Pursuit[R] over the soybean. Metolachlor was included to provide better coverage for some grasses and broadleaf species and also because it could be applied to the entire field each year.

1993 Flood Event

The 30-year average for total annual precipitation for Ames, Iowa, is 818 mm. Recorded precipitation totals in the WCW were 836 mm, 780 mm, 1,290 mm, 610 mm, and 700 mm, for calendar years 1991 through 1995, respectively. In only 1991 and 1993 were above-normal precipitation amounts recorded. Not surprisingly, largest deviations within the monthly totals relative to the 30-year averages were observed only in those years. In 1993, the deviations from the 30-year average for monthly precipitation in WCW was +49 mm, + 230 mm, and +152 mm, for June, July, and August, respectively (3). Describing events in terms of historical frequency of occurrence, 1993 was a "200-year flood" event. By late July and continuing into mid-August, much of the no-till field, with the exception of highest ridgetop locations, was underwater.

Field / Laboratory Methods

Paired piezometers were installed at fourteen locations distributed across the field reflecting different landscape positions and soil associations. Holes were drilled using a six inch hollow stem auger, with each pair spaced one meter apart. Each piezometer pair was assembled and each placed down-hole such that one meter of screened casing was positioned at the water table and at one meter below the water table. Piezometer screens penetrated depths ranging from 0.6 m at depressional areas to 2.4 m at ridgetop areas. Maximum water

Table III. Herbicide Application to No-Till East and No-Till West for Each Year of the Study.

Year	Herbicide	Rate	No-Till East	No-Till West
1992	metolachlor	2.0 kg / ha	Soybean	
	imazethapyr	0.07 kg / ha		
	bromoxynil	0.28 kg / ha		Corn
	atrazine	0.56 kg / ha		
	nicosulfuron	0.037 kg / ha		
1993	Glyphosate /	0.28 kg / ha	Corn	
	2,4-D	0.25 kg / ha		
	atrazine / cyanazine	1.09 kg / ha		
	clethodim	3.38 kg / ha		
	nicosulfuron	0.105 kg / ha cancelled -flood conditions		
	metolachlor	2.5 kg / ha		Soybean
	2,4-D	0.54 kg / ha		
	imazethapyr	0.07 kg / ha		
1994	metolachlor	2.0 kg / ha	Soybean	
	glyphosate	0.42 kg / ha		
	2,4-D	0.54 kg / ha		
	imazethapyr	0.07 kg / ha		
	atrazine /	1.09 kg / ha		Corn
	cyanazine	3.38 kg / ha		
	2,4-D	0.54 kg / ha		
	dicamba	0.14 kg / ha		
	2,4-D	0.13 kg / ha		
	nicosulfuron	0.037 kg / ha		
	2,4-D	1.12 kg / ha		
1995	atrazine /	1.09 kg / ha	Corn	
	cyanazine	3.38 kg / ha		
	nicosulfuron	0.035 kg / ha		
	2,4-D	1.08 kg / ha		
	metolachlor	2.18 kg / ha		Soybean
	2,4-D	0.53 kg / ha		
	glyphosate	0.42 kg / ha		
	imazethapyr	0.07 kg / ha		
	clethodim	0.105 kg / ha		

The following year both the cropping and the he post-emergence application was switched to the other half of the field. The herbicide application to each half of the field for each year is detailed in Table III.

table fluctuations were ± 2 m.The screened zone was packed in clean sand, the annular space back filled to the surface with bentonite clay, and the entire area surrounding the casing saturated with water. Groundwater samples were taken at both depths. Sampling sites were selected with traffic patterns strategically located so as to minimize the effects of localized traffic compaction on the piezometer response. Event related well water samples were collected. In the absence of rainfall, samples were taken at two-week intervals for the first six weeks following application and then monthly for the remainder of the growing season. An event was determined by the groundwater response as surmised from a change in water table level. If no event occurred, a sample was taken at the end of each month. Typically, well water samples were taken from May through November. All samples were collected manually using a TeflonR bailer with approximately one-liter capacity. Prior to sample collection, dissolved oxygen (DO) concentration was measured downhole using a commercial electrode sensor. Piezometers were then purged manually and sample withdrawn from fresh recharge within 24 hours. Four-liter samples were immediately directed through a flow cell configured for measurement of pH, specific conductance, and temperature, and then transferred to clean amber glass solvent bottles and held under cold storage in the field. To insure no contamination between each piezometer sampling, flow cell and transfer lines were rinsed with a water/methanol solution. All samples were returned to the laboratory on the day of collection where they were stored in the dark at 4 °C until extraction. A two-liter test sample was taken for herbicide residue analysis. Concurrently, a second measurement of pH, specific conductance, temperature and dissolved oxygen concentration, was taken in the laboratory. The residue analysis procedure for nicosulfuron and imazethapyr in groundwater was adapted from a USEPA-Industry protocol developed for ALS-inhibiting herbicides in surface water by a consortium of collaborators representing manufacturers (16-18). More recently, their approach has been modified to lower the detection limit by another order of magnitude for both surface and groundwater (19). The sample was filtered through a glass microfiber depth filter to remove sediment and acidified to pH=3 using dilute hydrochloric or acetic acid. The filtrate was taken to solid-phase extraction on RP-102 cartridges previously washed and activated with methanol and water. Following passage of the sample, the cartridge was eluted with acetone, taken to dryness, and reconstituted in 10% acetonitrile / 90% water acidified to pH= 3 with acetic acid. The final extract volume was adjusted to 1.0 ml. Instrumental analysis included separation by high performance liquid chromatography on reversed-phase n-octadecyl-silica packing using an acetonitrile / acetic acid / water-mobile phase. Identification and quantitation was performed using electrospray ioinization ion-trap mass spectrometry-mass spectrometry techniques.

Results / Discussion

Hydrology

Groundwater movement within the field is predominantly influenced by the tile drainage system. The east side of the field is drained by a tile line installed in the early 1900's, according to county tile map records. This tile drains the center section of the lower portion of the field, directly influencing water levels in piezometer 43, which is set within a pothole, and piezometer 41, which is set near the edge of the field (see Plates 1a and b). Rainwater infiltrated from the upslopes is stored in the ridges and migrates down gradient to the tile line. Surface runoff also migrates from the sideslopes and collects in the depressional areas. This tile line has no surface drain near the depressional area of piezometer 42. Water levels in piezometers 41, 42, and 43 vary the least of all locations on the east field.

Groundwater movement on the west field is influenced by three depressional areas; one near the center of the field, and one each near piezometers 48 and 49. A more recently installed surface tile drain is located approximately 200 meters north by northwest of piezometer 48, with the subsurface tile extending south to west of piezometer 48. Drainage for the west field moves to the west and north via the tile drains. Surface runoff pools at each of the depressional pothole areas on the west side but is removed more quickly via the surface tile drain. Saturated hydraulic conductivity (Ksat) was estimated for each well site using 1994 soil cores which were sectioned by soil horizon for laboratory measurements. These estimates are shown in Plate 2. Generally, the saturated soil on the upslopes and ridges shows greater Ksat (1 - 2 cm / hr) than the saturated soil in the depressional areas (Ksat < 0.25 cm / hr). For some locations a compaction difference is evident, presumably a consequence of previous tillage practices.

Water Quality Profile

When enhanced with long-term records of weather patterns and agronomic practices, the environmental impact at the field scale of a farming system can be defined in terms of a "water quality profile"(20,21). This conceptual model defines water quality in terms of a common suite of chemical properties together with the fate and distribution of agchemicals in each hydrogeologic compartment. The impact of the farming system on the environment can be quantified in terms of the cropping system, tillage practice, and fertilizer / pest management chemicals strategy. Several parameters were measured at both time of sampling in the field and again at time of extraction following laboratory storage. Piezometers 41, 43, 48, 49, and 51 showed DO concentrations which consistently fell between 0.4 - 2 mg / L. With the exception of piezometer 51, all are located near depressional areas. Piezometer

51 is located at a confluence point characterized by an east-west elevation change. Lower DO concentrations measured at these locations may be caused by the greater hydraulic head from groundwater in upgradient slopes. This could result in outgassing. DO concentrations at the highest elevations and sideslopes varied between 6 - 12 mg / L, usually increasing during periods of recharge from rainfall. Slug tests also confirmed the highest recharge rates at these sites. The deeper piezometer placement at each nest in the non-depressional areas consistently showed DO concentrations between 2 - 4 mg / L. This may be the result of oxygen utilizatization by heterotrophic anaerobes or by outgassing under the increased hydraulic head.

Field measurement of specific conductance of the groundwater at the time of sample collection showed some variation between depressional (0.7 - 1.0 $m\Omega$ / cm) and upslope (0.35 - 0.6 $m\Omega$ / cm) areas. This low conductivity could be an indication of the age of the water or the very slow rate of dissolution of mineral species in this relatively youthful glacial till. Water pH did not show a similar landscape position trend as did conductivity. Over a three-year period during transition to no-till, the pH of the groundwater dropped from nearly 8.0 to about 7.2 for the deeper piezometers. The shallower piezometer nests varied randomly between 7.0 and 7.7. Movement of applied fertilizer is most likely responsible for these variations, along with seasonal uptake of plant nutrients. Aqueous pH is important in the sorption process because it can greatly influence (a) the ionic nature of polar and amphoteric sorbates, thereby altering their binding properties, (b) the predominant charge strength or distribution on the surface of sorbents, and (c) water solubility of a sorbate thereby possibly influencing its bound concentration (22). Thus, changes in pH could alter environmental mobility by changing some physicochemical properties, such as soil distribution coefficient and water solubility. Concentration of nitrate varied across the field for all piezometer nests, ranging between 3 - 20 mg / L. The highest measured concentration of 20 mg / L was observed at piezometer 45-8 at a ridgetop position (see variation in hydraulic conductivity shown in Plate 2) while positions in depressional area nests 43, 48, and 41 were all measured at < 5 mg / L. During the three-year transition the nitrate levels increased slightly, while there appeared to be no correlation between concentration and depth. Excessive fertilization with anhydrous ammonia could account for a short-term increase in pH only to be effectively offset by increased rates of nitrification resulting in greater concentration of nitrate in the shallow ground water during transition. Consequently, pH decreases over time.

Herbicide Detections

Piezometer installation was completed in early winter of 1992. First application of Accent[R] and PursuitR occurred in Spring of that year. Groundwater samples taken in April 1993 revealed low concentrations of nicosulfuron (< 100 ng / L at 45-10) in the west field ridge and midslope position piezometers. The Pursuit[R] nicosulfuron degradate, J290, (2-amino-4,6-

dimethoxy 1,3-pyrimidine), was identified at somewhat lower concentrations (< 20 ng / L at wells 47-8, 48-10, and 46-8). These detections may be a consequence of higher percolation rate (see Plate 2) at site 46 with resulting transport down gradient in subsurface flow to reach sites 47 and 48. At piezometers 43-8 and 41-8, imazethapyr was detected at the shallower of the screened depths at 400 ng / L and 250 ng / L, respectively. Both are depressional area piezometers within the east field. These data indicate that some 74 cm of precipitation falling during the 9-10 months following application is sufficient to leach detectable concentrations of both herbicides to groundwater tables at 0.6-2.4 meter depths. Similarly, these could result from upgradient infiltration and movement under increased hydraulic head pressure. These occurrences partially result from the physicochemical properties of these two herbicides. Because of their structure, the acid imidazolinones possess unique acid-base properties. Any of five distinct chemical species may dominate in aqueous solution depending upon pH. In the pH range most critical for environmental issues (pH range 5-9) the species with the imidazolinone ring non-ionized but with the carboxylic acid group largely dissociated becomes dominant. As the pH drops from neutrality toward 5, the undissociated neutral species predominates. In contrast, nearly all sulfonylurea herbicides are weak acids, with acidities similar to acetic acid. Therefore, at pH's normally encountered in soils, pH 5-7, the sulfonylurea's are largely dissociated, with the dominant species being the anion formed by dissociation of the more acidic urea-hydrogen adjacent to the sulfonyl group on the sulfonylurea bridge. Depending on the buffering capacity of the soil water, the proportion of sulfonylurea anion increases as the pH rises. Also, as soil pH rises above 7, persistence appears to increase due to a decrease in the rate of hydrolysis. Another consequence of this weakly-acidic behavior is a pH-dependence on water solubility. For many sulfonylurea herbicides, water solubility increases by more than an order of magnitude when the pH is raised from 5 to 7 (10). Nicosulfuron water solubility increases from 390 mg / L to 18,000 mg / L when the pH is increased from 5 to 7. At pH 9, it becomes greater than 250,000 mg / L. The soil partition coefficient for imazethapyr is low and variable depending upon soil pH. At pH 7.0 and water solubility of 1,400 mg / L, calculated soil sorption coefficients range between 0.46 - 0.82 for sand- and silt-loam soils. Furthermore, residues in soil are reported to remain for periods up to three years following a single application, and this persistence seems to be enhanced even further in soils of higher clay and organic matter content (23,24). Calculations using data from aerobic soil metabolism studies indicate a half-life of 33 - 37 months. Thus, imazethapyr is both mobile and persistent. Similarly, the soil partition coefficient for nicosulfuron varies from 0.16 - 0.61 for sand- and silt-loams. However, it is much less persistent in soil than is imazethapyr, with typical half-life reported to be on the order of days; 26 - 63 days in silty clay loam soils.

As shown in Table III, Pursuit[R] was applied to the west field on July 12, 1993. Analysis of piezometer samples collected on July 28, 1993, following 14 cm of rainfall revealed the presence of imazethapyr in wells 47-9 and 46-8 at 1.3 µg / L and 230 ng / L, respectively. Nicosulfuron and J290 (2-amino-4,6-

Table IV. Summarized Concentration Data from Across the Field for Nicosulfuron, J290, and Imazethapyr for 1993-1995*.

Analyte	Average Concentration	Depth	Percentage Detections	Median Concentration**
Nicosulfuron	131 ng / L	< 3 m	16%	96 ng / L
		3 m	15%	169 ng / L
J290	24 ng / L	< 3 m	24%	38 ng / L
		3 m	22%	16 ng / L
Imazethapyr	125 ng / L	< 3 m	33%	136 ng / L
		3 m	24%	102 ng / L

* sample dates: 01/1993......04/1993, 01/1994......10/1995; total samples analyzed = 250.

** median concentration segregated by depth due to skewedness of distributions

dimethoxy-1,3-pyrimidine) were not detected at this time, which would suggest that both were either degraded to intermediates beyond J290 during the 16 days since application or were diluted to levels below our detection limits by the heavy rainfall. Due to the flooding, Accent^R was not applied to the east field during 1993.

Table IV provides a summary for all detections across the field during 1993-1995 for nicosulfuron, J290, and imazethapyr.

Mean and median concentrations are given along with the percentage distribution for detects above and below the three-meter depth. In addition, statistics are offered for confirmation of identity by mass spectrometry and mass spectrometry-mass spectrometry. In order to emphasize positional effects within the field, groundwater data from all 1993-1995 sampling is compiled and presented as a summation of detections and concentrations relative to a piezometer location. Plates 3a, 3b, and 3c illustrate these summations for nicosulfuron, J290, and imazethapyr, respectively. This is done in order to observe site specific trends, as the high water solubility and rapid flux in the near surface water table results in very transient time series events. This rapidly changing time series is a consequence of both the kinetics of degradation (specifically hydrolysis, and more important for nicosulfuron) together with the changing amphoteric character of parent (more important for imazethapyr) under an environment of changing pH. As shown in Plate 3a, nicosulfuron detections were frequently associated with piezometers which intercept gradient flow to the tiles or were located in or near surface depressions. These positions have the greatest likelihood of receiving water carrying a herbicide load due to their ability to intercept and conduct water to the tile lines. An exception is seen for piezometers 39-7 and 39-10 located on the east ridge top where 1995 detections for nicosulfuron ranged from 2.6 μg / L following application to as low as 76 ng / L over the succeeding five months. As seen in Plate 2, site 39 shows greater hydraulic conductivity consistent with this type of observation. Nevertheless, nicosulfuron has moved through the soil profile and leached to the water table, an observation in disagreement with published laboratory studies suggesting very low mobility through loamy and clayey Iowa soils due to rapid chemisorption onto 2:1 clay minerals (25). Changes in groundwater chemistry may be much slower at this location, or, alternatively, this area may be a point at which boom sprayer overlap resulted in effectively doubling the application. Plate 3b shows the weighting of detections over the period for J290, which is not detected in higher concentration than nicosulfuron, from which it is derived. The east half of the field received only a single application of Accent[R] over the period 1992-1995; whereas, the west half of the field received two applications. Plate 3b confirms that more J290 residue is reaching the groundwater on the west half. The ridge location at 45-8 gave one extremely high detection, 885 ng / L, which coincided with a nicosulfuron detection of 2 µg / L on June 7, 1994, only 10 days following application. Both detections are very transient as neither are detected only one week later on June 15 following approximately 4 cm of rainfall. These type of results propound the complexity of sampling the near surface water table for very hydrophilic and water soluble constituents like nicosulfuron and J290 which are not substantially sorbed to the soil matrix. Generally, Table IV shows that concentrations of nicosulfuron reaching shallow water tables is very low overall and that deeper piezometer of each pair frequently reveals the higher loadings. For J290, the biodegradation may proceed more slowly than for nicosulfuron resulting in higher concentrations at the near-surface piezometer locations. Comparing Plates 3a and 3b it is apparent that J290 is more consistent in the number of observations per site than the parent herbicide.

Summation of detection frequency for imazethapyr across the field is illustrated in Plate 3c. Most of these are at depressional locations where groundwater movement is dominated by tile drainage. Table IV shows the frequency of detection for imazethapyr to be almost twice that for nicosulfuron, affirming its very rapid movement to shallow water tables. Other detections are not as easily explained. For example, 2.3 µg / L was measured at location 39-8 on July 7, 1995, following an application on June 19, 1995. Similarly, at site 46-8 the concentration of imazethapyr consistently increased to 200 ng / L post application in both 1993 and 1995 (see Plate 2). Macropore flow resulting from any combination of earthworm movements, plant root channels, or channeling from the surface due to faulty piezometer installation all may account for these occurrences.

The residue analytical methodology employed in this study was not limited to detection and quantification of nicosulfuron and imazethapyr. Other families of herbicides were also included in the screening; specifically, the sym-triazines and the chloroacetanilides. Over the three year effort, atrazine, cyanazine, metolachlor, alachlor, acetochlor and several degradates all were detected in various piezometers at concentrations ranging from 10 µg / L to 10 ng / L. The 1993 results for the above-mentioned chemicals stand as evidence that herbicides from application on other nearby fields may have somewhat skewed some of the experimental results potentially leading to incorrect interpretation of the environmental impacts of no-till farming practices. However, it should be pointed out that these other herbicide families represent the same chemicals which were used repeatedly on this landscape prior to our study. Concomitantly, imazethapyr and nicosulfuron were applied in 1992-95 and no detections prevailed without short-term secession of detections in the samples. No samples were collected during post-flood conditions in 1993. February 1994 samples yielded only three detections of imazethapyr, all < 20 ng / L. Results are validated for all three years of monitoring by examination of the 1994 1995 data, which shows the behavior of these chemicals independent of the flood conditions of 1993.

Residence time for initial detection in shallow groundwater was 24-88 days and 30-60 days for imazethapyr and nicosulfuron, respectively. In order to enhance the understanding of the component processes underlying their environmental fate in our system, increased sampling intervals are needed, in addition to our event-based sampling design. With more intensive sampling of the soil, complete profile distributions could be obtained, leading to better estimates of degradation rates and transport functions from residence times. This approach would also be necessary for accurately estimating load or flux through a prescribed soil volume. Sampling shallow groundwater at increased frequency would require smaller sample volumes in order to avoid excessive perturbation of the groundwater unnecessarily by recharge within the vicinity of the sampling region. Accurate and reproducible results for imazethapyr and nicosulfuron in water can be achieved by utilizing ELISA techniques for quantitative screening on very small sample volumes (26). Unfortunately, these kits are no longer available and would require custom development and manufacture. Such services are commercially available; however, the costs are high.

When comparing soil translocation time, it becomes apparent that it is much faster for the sulfonylurea and imidazolinone herbicides than it is for the triazine and chloroacetanilide herbicides, even though they are frequently applied at only a fraction of the rates. This study also confirms that using nested sites above shallow ground water tables on tile-drained landscapes is an insightful field design wherein both hydrologic and biogeochemical processes can be studied in some detail. The mobilities of these compounds indicate they are likely to be found in surface water as well, as tile drainage emerges as surface flow. This was confirmed by two noteworthy reconnaissance studies conducted across the Midwest within the past three years. The first was a collaborative effort among the U.S. Geological Survey, the Agricultural Research Service, and the duPont Company (16). The second, and more recent, was a similar and expanded effort involving the U.S. Geological Survey and the duPont Company (27). Both studies confirm that the chemicals are highly mobile through the water resources normally associated with rainfed production agriculture in the Midwest and demonstrate the basic need for aggressive monitoring in the years ahead in order to mitigate the longterm effects of their pollution potential.

Summary

Implementing a post-emergence herbicide-based weed control strategy during transition from chisel plow conventional tillage to no-tillage under typical Iowa rainfall patterns results in changes to shallow groundwater quality. Following only 1-3 years of application of both Accent[R] and Pursuit[R], the active ingredient in each formulation is detected in shallow groundwater beneath Des Moines Lobe glacial till soils. In these largely tile-drained systems, frequency of detections and range of concentrations are governed mostly by rainfall patterns and subsurface hydrology. While this landscape is generally of low relief, differences in groundwater chemistry between ridge and pothole locations are discernable. Results suggest that new generation chemistries, such as those represented by nicosulfuron and imazethapyr, must be included in future monitoring programs if our ground water resources are to be effectively managed and protected. Results also confirm the importance of including those degradates, rapidly produced by the biogeochemistry occurring within surface soil horizons, which are likely to persist as a consequence of their properties. Definitive understanding of the routes of soil dissipation and degradation of an ever increasing number and complexity of new herbicides is necessary in order to conserve and preserve the fragile soil landscape on which the U.S. agricultural abundance depends (28).

References

1. Hatfield, J. L., Bucks, D. A., and Horton, M. L. In *Agrochemical Fate and Movement Perspective and Scale of Study*; Steinheimer, T. R, Ross, L. J., and Spittler, T. D., Eds.; ACS Symposium Series 751; American Chemical Society: Washington, DC, 2000, Chapter 16; pp232-247.
2. Bucks, D. A., *Agric. Res.*, September, **1993**; pages 2-11.
3. Hatfield, J. L., Jaynes, D. B., Burkart, M. R., Cambardella, C. A., Moorman, T. B., Prueger, J. H., and Smith, M. A., *J. Environ. Qual.*, **1999**; *28*, 11-24.
4. Cambardella, C. A., Moorman, T. B., Novak, J. M., Parkin, T. B., Karlen, D. L., Turco, R. F., and Konopka, A. E., *Soil Sci. Soc. Am. J.*, **1994**; *58*, 1501-1511.

5. Andrews, W. F. and Dideriksen, R. O. 1981. *Soil Survey of Boone County, Iowa*. U. S. Government Printing   Office, Washington, D. C.

6. Eidem, J. M., Simpkins, W. W. and Burkart, M. B., *J. Environ. Qual.*, **1999**; 28, 60-69.

7. Sauer, P. A. and Hatfield, J. L., *Walnut Creek Watershed Research Protocol Report*, 1994; Bulletin Number 94-1, U. S. Dept. Of Agriculture, Washington, D. C. [report available upon request from the USDA-ARS, 2150 Pammel Drive, Ames, IA., 50011].

8. Lal, R., Logan, T. J., Eckert, D. J., Dick, W. A. and Shipitalo, M. In *Conservation Tillage in Temperate Agroecosystems*, Carter, M. R., Ed., CRC Press, Inc., Boca Raton, FL., 1994; Chapter 4, pp73-114.

9. Uri, N. D., *J. Sustain. Agric.*, **1999**, 15(2-3), pp 5-17.

10. Beyer, E. M., Jr., Duffy, M. J., Hay, J. V. and Schlueter, D. D. In *Herbicides: Chemistry, Degradation, and Mode of Action*, Volume 3, Kearney, P. C. and Kaufman, D. D., Eds., Marcel Dekker, Inc., New York, NY, 1988, Chapter 3, pp 117-189.

11. Shaner, D. L. and O'Connor, S. L. In *The Imidazolinone Herbicides*, CRC Press, Inc., Boca Raton, FL, 1991, p 290.

12. *Agricultural Chemical Usage, 1992 Field Crops Summary*, prepared by the National Agricultural Statistics Service, U.S. Government Printing Office: Washington, D.C., 1993, 118 pp.

13. *Agricultural Chemical Usage, 1993 Field Crops Summary*, prepared by the National Agricultural Statistics Service, U.S. Government Printing Office: Washington, D.C., March 1994, 114 pp.

14. *Agricultural Chemical Usage, 1994 Field Crops Summary*, prepared by the National Agricultural Statistics Service, U.S. Government Printing Office: Washington, D.C., March 1995, 106 pp.

15. *Agricultural Chemical Usage, 1995 Field Crops Summary*, prepared by the National Agricultural Statistics Service, U.S. Government Printing Office: Washington, D.C., March 1996, 100 pp.

16. Steinheimer, T. R., Pfeiffer, R. L., Scoggin, K. D., and Battaglin, W. A. In *Agrochemical Fate and Movement-Perspective and Scale of Study*; Steinheimer, T. R, Ross, L. J., and Spittler, T. D., Eds.; ACS Symposium Series 751; American Chemical Society: Washington, DC, 2000, Chapter 17, pp 248-271.

17. Rodriguez, M. and Orescan, D. B., *Anal. Chem.*, **1998**, 70(13), pp 2710-2717.

18. Krynitsky, A. J., *J. AOAC Int.*, **1997**, 80(2), pp 392-400.

19. Furlong, E. T., Burkhardt, M. R., Gates, P. M., Werner, S. L., and Battaglin, W. A., *Sci. Total Environ.*, **2000**, 248(2-3), pp 135-146.

20. Steinheimer, T. R., Scoggin, K. D., and Kramer, L. A., *Environ. Sci. Technol.*, **1998**, *32,(8)*, pp 1039-1047; *ibid.*, pp 1048-1052.

21. Steinheimer, T. R. and Scoggin, K. D., *J. Environ. Monit.*, **2001**, 3, 126-132.

22. Koskinen, W. C. and Harper, S. S. In *Pesticides in the Soil Environment: Processes, Impacts, and Modeling*, Editor, H. H. Cheng, SSSA Book Series: 2, Soil Science Society of America, Inc., Madison, WI., 1990, pages 51-79.
23. Loux, M. M., Liebl, R. A., and Slife, F. W., **1989**, *Weed Science*, 37, 259-267.
24. Loux, M. M. and Reese, K. D., *Weed Technol.*, **1993**, 7, 452-458.
25. Gonzalez, H. and Ukrainczyk, L., *J. Environ. Qual.*, **1999**, 28, 101-107.
26. Steinheimer, T. R. and Scoggin, K. D., *J. Agric. Food Chem.*, **1998**, 46(5), 1883-1886.
27. Battaglin, W. A., Furlong, E. T., Burkhardt, M. R. and Peter, C. J., *Sci. Total Environ.*, **2000**, 248(2-3),   123-133.
28. Steinheimer, T. R., Ross, L. J., and Spittler, T. D. In *Agrochemical Fate and Movement Perspective and Scale of Study*, Steinheimer, T. R, Ross, L. J., and Spittler, T. D., Eds.; ACS Symposium Series 751; American Chemical Society: Washington, DC, 2000, Chapter 1, pages 2-19.

Chapter 16

Photostability of Two Fungicides on Spray Application Monitors: Effect of Paper and Formulation Type

Joseph H. Massey[1] and Suzanne Koch Singles[2]

[1]Mississippi State University, Box 9555, 117 Dorman Hall, Mississippi State, MS 39762
[2]Stine-Haskell Research Center, DuPont Agricultural Products, P.O. Box 30, Newark, DE 19711

Application monitors are used to determine the accuracy and precision of pesticide applications made for terrestrial field dissipation studies. Our results show that the photostabilities of two fungicides were affected by the type of paper used as the application monitor and by pesticide formulation. In side-by-side comparisons using an emulsifiable concentrate (EC) formulation, rapid photodegradation of both fungicides occurred when applied to white $\propto$-cellulose paper. Only 10 to 35% of the applied ^{14}C-fungicides remained as parent after 15-min irradiation in a Suntest photolysis chamber. Dark controls exhibited no losses or degradation under study conditions. In contrast, greater than 85% of the compounds remained as parent when applied to brown Kraft paper. These effects were attributed, in part, to enhanced scattering of light by the white background of the $\propto$-cellulose paper. When applied as a suspension concentrate (SC) formulation, both fungicides were stable to photolysis on $\propto$-cellulose and Kraft papers. Differences between these formulations were attributed to either the presence of a photosensitizing agent in the EC and/or increased photostability of the active ingredients in the SC solid phase.

Introduction

A variety of application monitors has been used to assess the accuracy and precision of pesticide applications made in terrestrial field dissipation studies. Common types of monitors include petri dishes, paper or polyurethane-foam disks and soil-filled containers of various size (*1*). A key benefit of using application monitors is that variability associated with test substance application can be determined independent of errors typically associated with the collection, processing and analysis of soil cores.

Prior to use of a given application monitor in the field, it is important to assess test substance stability on, and recovery from, the monitor. The objectives of this work were to (A) develop a laboratory method to apply pesticides to application monitors that simulates field application deposition patterns and spray volumes, (B) develop a laboratory procedure to assess the photostability of pesticides on application monitors, allowing for the trapping of potential volatile products, and (C) determine the effects of paper type and formulation on the photostability of two photo-labile fungicides.

Materials and Methods

Paper Types: The two paper types evaluated were ∝-cellulose (Schleicher and Schuell, Inc.; Keene, NH) and Kraft paper (Fulton Paper Co.; Wilmington, DE). Squares (10- x 10-cm) were used for all studies.

Formulation Types: An emulsifiable concentrate (EC) formulation blank was used to prepare solutions containing either radiolabeled Fungicide A or Fungicide B.[1]
   A suspension concentrate (SC) containing 225 g/L of Fungicide B plus 75 g/L Fungicide A was used to prepare non-radiolabeled test solutions containing both active ingredients as a 30% SC (w/v).

Test Solution Preparation: For the EC formulations, 0.100 mL of EC formulation blank plus 0.200 mL of either ^{14}C-Fungicide A (98% radiochemical purity, specific activity = 68.3 µCi/mg) or ^{14}C-Fungicide A (94% radiochemical purity, specific activity = 10.0 µCi/mg) stock solution were added to a 5-mL volumetric flask and diluted to volume with deionized water. The ^{14}C-Fungicide A and ^{14}C-Fungicide B stock solutions contained approximately 1 mg a.i./mL dissolved in acetonitrile (ACN).

---

[1] Because these compounds have not been registered, their chemical structures and identities will not be divulged at this time.

For the SC formulations, 0.667 mL of 30 SC formulation was added to a 100-mL volumetric flask and diluted to volume with deionized water. This resulted in a solution concentration of 0.5 g Fungicide A + 1.5 g Fungicide B per liter.

Application Method: A DeVilbiss atomizer (Model 152, Sunrise Medical, Somerset, PA) attached to the in-house nitrogen gas supply was used to apply a fine mist of test solution to the application monitors. Prior to application, tare weights of the application monitors were measured and recorded. Applications began by placing three monitors on the bottom of a large paper bag in which a slit had been cut. The slit, located approximately 10 cm from the bottom of the bag, allow the atomizer to be moved across the entire length of the bag. Preliminary work showed three passes with the atomizer were needed to yield the desired application volume of 0.2 to 0.3 grams per monitor (equivalent to an application volume of 200 to 300 L/ha). During application, the paper application bag remained under a fume hood.

Immediately after application, the monitors were re-weighed to determine the amount of test solution deposited onto each monitor. The difference between the weights of each monitor before and after application was used to determine the actual volume of test solution applied to each monitor. A test solution density of 1.00 g/mL was assumed. Using this technique, three spray monitors could be treated at one time: one treated monitor served as a dark control while the other two monitors were used for the irradiated treatments.

Exposure of Application Monitors to Simulated Sunlight: Immediately after weighing the monitors treated with the EC formulation, two were placed into separate 20 x 20-cm Plexiglas chambers that were covered with quartz lids and fitted with sampling ports to allow constant flow of air through the chambers. Gas traps were used to collect potential volatile products. The exiting air stream was passed through 25-mL methanol followed by 25-mL sodium hydroxide. Temperatures of the lighted and dark-control chambers were maintained at approximately $25 \pm 2°C$ by circulating chilled water through heat exchangers located along the bottom of the Plexiglas chambers.

The Plexiglas chambers were positioned under a Heraeus CPS Accelerated Suntest unit (Heraeus Instruments; Phoenix, AZ) that uses filters to mimic natural sunlight (i.e. wavelengths ≥290 nm). The spectral irradiance of the Suntest unit approximated the intensity of mid-day Delaware summer sunlight (approximately 29 W/m^2 for 300 to 385 nm). Once the apparatus and samples were ready, the treated application monitors were irradiated for 15 minutes.

For the SC treatments, the above procedures were used with the exception of the gas traps. Radiolabeled test materials were not used with these treatments.

Extraction and Analysis of Application Monitors: The dark-control and irradiated monitors were placed into separate 250-mL Erlenmeyer flasks and extracted three times with 100-mL ethyl acetate. During each extraction the application monitors were allowed to soak for 30 minutes followed by hand swirling for one minute. To remove any potential polar products, the monitors were further extracted three times with 100-mL 9:1 acetonitrile:water. Ethyl acetate and ACN extracts for each monitor were combined and evaporated to dryness using a Turbo Vap ZW-500 at 40°C (Zymark; Hopkinton, MA). Extracts were then dissolved in 5-mL of 50:50 acetonitrile:water.

For studies involving radiolabeled materials (EC formulation only), liquid scintillation counting techniques were used to measure radioactivity in the HPLC fractions collected during HPLC analysis of the radiolabeled samples. For the SC formulation studies, only UV detection was used. For all samples, the extracts were injected directly onto a HPLC using the conditions given below:

HPLC System: Waters Alliance Series, Model 2690 Separations module, Waters photodiode array UV detector model 996, Waters Bus SAT/IN module, Millennium software ver. 2.15 (Waters; Milford, MA).

Column: Zorbax® SB-C3 column (150 mm x 4.6 mm, 5-μm particles)
Flow rate: 1.0 mL/min
Injection volume: 200 μL
Detection wavelength: 254 nm
Fraction collector: Isco Foxy (1-min fractions)
HPLC Gradient Program:

Time (min)	Solvent A (%)[1]	Solvent B (%)[2]
0	20	80
50	90	10
52	100	0
55	20	80

[1] A = Acetonitrile
[2] B = 0.1% Formic Acid

Combustion Analysis of Application Monitors with EC formulation: After extraction, each application monitor was mechanically pulverized using a Tissumizer (Tekmar Co.; Cincinnati, OH). Next, the monitors were subsampled (3 x 1gram) for combustion analysis on a Tri-Carb 306 Oxidizer (Packard Instruments; Downers Grove, IL) to determine unextracted residues. This was only performed on the application monitors containing radiolabeled materials.

Results and Discussion

The procedures developed in this study allow for the determination of test substance stability on pesticide application monitors using realistic spray deposition patterns and application volumes. Although spray droplet size was not actually measured, the application technique using the DeVilbiss atomizer visually appeared to approximate the deposition of fine droplets that occurs with field application equipment. Laboratory applications made to application monitors via mechanical pipet, on the other hand, would not have easily duplicated the deposition of these fine droplets. In addition, the application procedure was fairly reproducible (CV = 33%) with application amounts averaging 0.264 ± 0.0867g test solution per monitor (n = 20). Assuming a test solution density of 1.00 g/mL, the 0.264 g/monitor corresponds to an application volume of 264 L/ha. Ultimately, homogenization and combustion of the application monitors represented the most troublesome aspects of this procedure.

The mass balances for Fungicide A ranged from 94.9% to 120.0% of applied radioactivity (AR) with volatiles accounting for <1% AR and unextracted (bound) residues accounting for <1% to 8.0% AR for the three treatments (Table I). When applied as the 10 EC formulation to the α-cellulose monitors, Fungicide A (31 min) degraded to photoproduct A (24 min) after 15 minutes of irradiation (Figure 1). In contrast, Fungicide A was stable when applied to Kraft paper monitors. The difference between paper types may be due to enhanced scattering of light by the white background of the α-cellulose monitor, resulting in increased photodegradation of Fungicide A. Light clearly played a role in this result since the test substance was stable on the α-cellulose monitor not exposed to light (Figure 1).

The mass balances for Fungicide B ranged from 81.4% to 129.0% AR with volatiles again accounting for <1% AR (Table II). These mass balances reflect some of the variability problems we experienced with the homogenization process. When applied as the 10 EC formulation to the α-cellulose monitors, Fungicide B (22 min) degraded to photoproduct B (19 min) after 15 minutes irradiation (Figure 2). In contrast, Fungicide B was relatively stable when applied to brown Kraft paper monitors. This result may also be due to enhanced scattering of light by the white background of α-cellulose since Fungicide B was stable on the α-cellulose monitor kept in the dark (Figure 2).

When applied as the 30 SC formulation, Fungicide A (29.6 min) appeared to be stable on both α-cellulose (Figure 3) and Kraft-paper (Figure 4) monitors. Unlike the EC formulation, no Fungicide A photoproduct (23.0 min) was detected when the SC formulation was used. Photoproduct B (22.6 min), the photoisomer of Fungicide B (24.8 min), was formed on the α-cellulose paper

Table I. Mass Balance Results for ^{14}C-Fungicide A Applied as an EC Formulation to Two Application Monitors.

Application Monitor Treatment	Tare Weight (g)[1]	Weight After Application (g)[2]	Spray Solution Applied (g)[3]	Total DPM Applied	Extracted Residues (% AR)[4]	Bound Residues (% AR)[5]	Volatiles (% AR)[6]	Mass Balance (% AR)
α-cellulose irradiated	10.338	10.520	0.182	268990	86.6	8.0	0.3	94.9
Kraft paper irradiated	3.972	4.104	0.132	186500	114.4	5.5	0.1	120.0
α-cellulose Dark-Control	9.957	10.232	0.275	313500	105.1	0.6	0.0	105.7

[1]Tare weight of application monitor before application of test solution.

[2]Weight of application monitor after application of test solution.

[3]Weight of test solution applied used to estimated volume of test solution applied (assumes test solution density of 1.00 g/mL).

[4]Combined percentage of Applied Radioactivity (AR) recovered from ethyl acetate and 9:1 ACN:water extractions.

[5]Determined by post-extraction combustion analyses of homogenized monitor.

[6]Combined percentages of AR found in methanol and ethylene glycol traps.

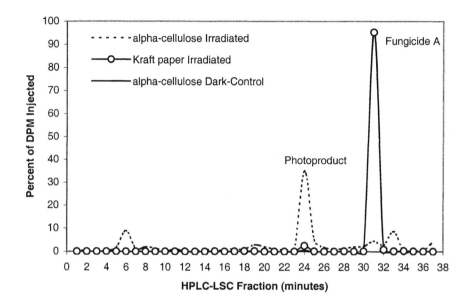

Treatment	Fungicide A (% AR)	Photoproduct of A (% AR)
α-cellulose irradiated	9.2	37.5
Kraft paper irradiated	87.1	< 1
α-cellulose Dark-Control	98.5	< 1

Figure 1. Effect of Paper Type on the Photostability of Fungicide A (as EC Formulation) on Two Application Monitors after 15-min Irradiation.

Table II. Mass Balance Results for ^{14}C-Fungicide B Applied as an EC Formulation to Two Application Monitors.

Application Monitor Treatment	Tare Weight (g)[1]	Weight After Application (g)[2]	Spray Solution Applied (g)[3]	Total DPM Applied	Extracted Residues (% AR)[4]	Bound Residues (% AR)[5]	Volatiles (% AR)[6]	Mass Balance (% AR)
α-cellulose irradiated	10.274	10.603	0.329	1378252	85.0	8.8	0.0	93.8
Kraft paper irradiated	4.599	4.720	0.121	502667	55.9	73.1	0.0	129.0
α-cellulose Dark-Control	10.167	10.436	0.269	1123780	77.1	4.3	0.0	81.4

[1]Tare weight of application monitor before application of test solution.

[2]Weight of application monitor after application of test solution.

[3]Weight of test solution applied used to estimated volume of test solution applied (assumes test solution density of 1.00 g/mL).

[4]Combined percentage of Applied Radioactivity (AR) recovered from ethyl acetate and 9:1 ACN:water extractions.

[5]Determined by post-extraction combustion analyses of homogenized monitor.

[6]Combined percentages of AR found in methanol and ethylene glycol traps.

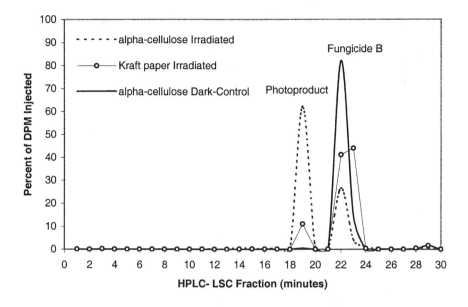

Treatment	Fungicide B (% AR)	Photoproduct of B (% AR)
α-cellulose irradiated	31.3	61.6
Kraft paper irradiated	84.6	10.2
α-cellulose Dark-Control	96.8	< 2

Figure 2. Effect of Paper Type on the Photostability of Fungicide B (as EC Formulation) on Two Application Monitors after 15-min Irradiation.

(Figure 3) and, to a lesser extent, on Kraft paper (Figure 4). These results suggest both Fungicide A and Fungicide B were most stable when the SC was applied to the Kraft paper monitors. As before, both active ingredients were stable on the dark control (Figure 5). These results are summarized in Table III.

Conclusions

The test substance application and exposure techniques developed in this study can be used to determine the photostability of a test substance on an application monitor using realistic application volumes and spray deposition

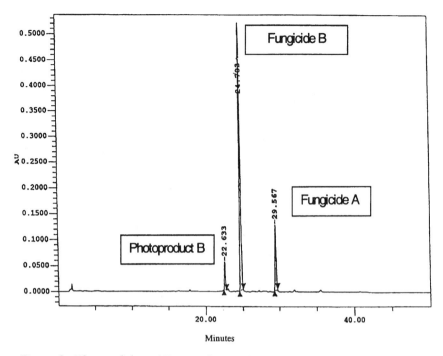

Figure 3. Photostability of Fungicide A and Fungicide B (as SC Formulation) on α-Cellulose Paper Application Monitor after 15-min Irradiation.

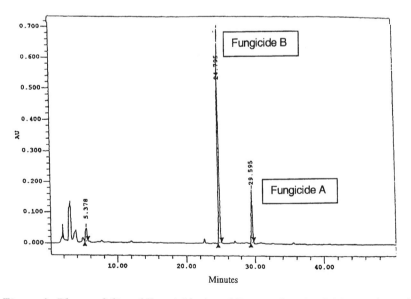

Figure 4. Photostability of Fungicide A and Fungicide B (as SC Formulation) on Kraft Paper Application Monitor after 15-min Irradiation.

Table III. Summary of Results for Fungicide A and Fungicide B Applied as an SC Formulation to Two Application Monitors.

Application Monitor Treatment	Test Solution Applied (g)	Peak Areas				
		Photoproduct B (R_t = 22.6 min)	Photoproduct A (R_t = 23.0 min)	Fungicide B (R_t = 24.8 min)	Fungicide A (R_t = 29.6 min)
α-cellulose irradiated	0.3320	501729	not detected	5069442	1390136
Kraft paper Irradiated	0.2958	small peak detected	not detected	7035211	1930787
α-cellulose Dark-Control	0.3254	not detected	not detected	6154550	1036182

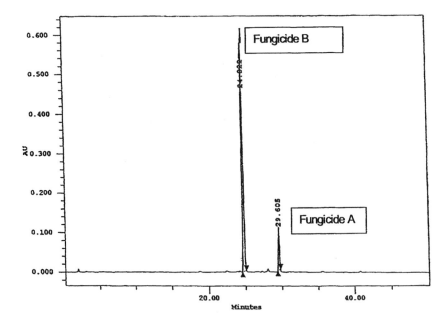

Figure 5. Stability of Fungicide A and Fungicide B formulated (as SC Formulation) on α-Cellulose Application Monitor (Dark Control).

patterns. Formulation (EC vs. SC) and paper type (α-cellulose vs. Kraft paper) significantly affected the photostabilities of both fungicides. To reduce the potential for photodegradation in the field, application monitors made of Kraft paper may be a better choice than α-cellulose paper for photolabile compounds. Also, it is always prudent to collect and store application monitors from the field as soon as possible after test substance application. These results confirm the value in determining the stability and recovery of a test substance on a given application monitor prior to using the monitor in a terrestrial field dissipation study.

Reference

1. American Crop Protection Association. Workshop minutes on "Zero-time residue levels in field soil dissipation studies." 1996, Crystal City, VA. (19 Sept 1996)

Chapter 17

Experiments in a Volatilization Chamber under Simulated Outdoor Conditions: A Contribution to a Better Understanding of Field Dissipation Studies

R. Kubiak

State Institution for Research and Training in Agriculture, Viticulture and Horticulture, Ecology Department, Breitenweg 71, D–67435 Neustadt, Germany

An experimental volatilization chamber is presented which allows for full mass balance studies after application of ^{14}C-labeled compounds to plants and/or soils. The design of this chamber allows for the control of environmental parameters which influence the metabolism and volatilization of pesticides; either standardized climatic conditions, or measured and recorded outdoor scenarios can be used. This paper provides the results of chamber studies using the pesticides fenpropimorph and parathion-methyl, these results illustrate the reproducibility, exactness and with the example of parathion methyl the transferability of the data to field studies.

Introduction

After the application of a pesticide in agriculture, a substantial fraction of the dosage applied may enter the atmosphere and may be transported over long distances (1). The rate and the extent of the emission after application depends on the physical and chemical properties of the pesticide, the application parameters, the climatic conditions during and after application as well as the characteristics of the target. Vapor pressure and Henry's law constant as well as the kind of formulation used (2) are important parameters of the plant protection agent, and the droplet size as well as the water amount are application characteristics to be taken into account. Furthermore, it is well known that volatilization processes may be influenced by the relative humidity (3), air temperature (4), atmospheric pressure and wind velocity (5). Furthermore, irradiation may influence the metabolism of compounds via direct or indirect photolysis (6). Last but not least the sorption and desorption processes as well as transport to deeper soil layers and chemical and biological degradation in the soil may depend not only on soil characteristics such as structure, humus content and biological activity but also on soil temperature, soil moisture and pH-value (7). Since plants have fewer sorption sites than soils, volatilization of applied compounds is normally higher from plant than from soil surfaces but also on plant surfaces the potential amount of volatilization may be reduced by uptake into the leaves or degradation processes on the plant surface (8). In field dissipation studies, especially the climatic conditions vary during the experimental time and between different experiments which makes field dissipation irreproducible. Furthermore, field studies are open experimental systems, i.e. a distinction between degradation processes, the formation of bound residues and volatilization of the pesticide sprayed or any metabolites formed is not possible.

To contribute to a better interpretation of field dissipation studies a laboratory chamber was designed fulfilling the following requirements:

1. The use of formulated plant protection agents with the option of using ^{14}C-labeled active ingredients
2. An application procedure, which is in accordance with agricultural practice as regards nozzle type, water amount ha[-1], spraying pressure and target distance
3. An experimental chamber in which the climatic parameters can be simulated that may influence the fate of the compound including volatilization
4. A direct measurement of the volatile compound and the possibility of obtaining mass balances when using ^{14}C-labeled compounds.

To investigate the precision and the reproducibility of the results obtained from the chamber, experiments were carried out with different ^{14}C-labeled compounds. To investigate the transferability of the results obtained from these chamber experiments to the field, a comparative experiment was carried out with parathion-methyl.

Material and Methods

Design of the Application Chamber and Application Procedure

In order to ensure well-controlled application conditions a computer-controlled spraying system was built which enables application amounts between 100 and 2000 L ha[-1], spraying pressures between 1.0 and 3.0 bar and driving speeds between 0.1 and 30 km h[-1]. Nozzles licensed for agricultural use can be connected to this system in such a way that only the nozzle itself reaches into the application chamber. This chamber consists of stainless steel, can be closed with a door and an experimental platform containing bare soil or a plant/soil system can be introduced. The size of the experimental area is 0.5 m[2]. It is fixed on a platform of variable height. Depending on the experiment, bare soil or plant stands of different heights can be sprayed with the correct distance between the nozzle and the target area (9). When using ^{14}C-labeled active ingredients application losses can be determined precisely by measuring the radioactivity remaining in the spray system and at the chamber walls which are covered with filter paper during application. Figure 1 shows the equipment.

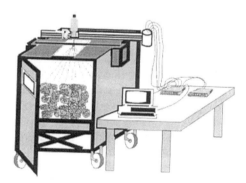

Figure 1. Application chamber for simulation of field application scenarios under controlled conditions. (Reproduced with permission of reference 10. Copyright 1999 Verlag Dr. Köster)

Before application aliquots of the homogenized radioactive application solution are measured using a liquid scintillation counter (LSC) and the total amount of radioactivity is calculated. A few minutes after application all fine droplets have reached the target and the chamber door is opened. The experimental platform removed for installation in the experimental chamber. The residual volume of the solution in the spraying equipment is determined and the whole spraying system is extracted with a suitable organic solvent. The remaining radioactivity is determined by measuring aliquots in a LSC. The paper, that covers the inner walls of the application chamber during application is removed and extracted with a suitable solvent. The volume is determined and aliquots are measured in a LSC. The total radioactivity which has not reached the target is calculated from these measurements and the amount is deducted from the total radioactivity measured in the spraying solution. The balance is the amount of ^{14}C which has reached the target. This very precise result is the 100% basis for the mass balance of the experiment. Knowing the specific radioactivity of the labeled compound (kBq mg^{-1}) and the percentage-value of the active ingredient in the formulation the exact amounts of radio-labeled compound and of formulated product applied can be calculated.

Design of the Volatilization Chamber and its Possibilities

After application, the experimental platform is introduced into the volatilization chamber. This working step lasts only 1 to 2 minutes, thus avoiding considerable ^{14}C-losses via volatilization. At the bottom of the platform (5 cm deep for experiments with bare soil and 10 cm deep for experiments with plants) an irrigation system made of perforated tubes is installed and connected to a water tank and a pump outside the chamber to control soil moisture during the experiments.

The chamber equipment consists of an air conditioning system, suitable for the on-line production of air temperatures and humidities ranging from 10 to 30 °C and 35 to 95% rh, respectively. The conditioned air reaches an equalization chamber, where the turbulence caused by the transport blowers is calmed down and converted into two parallel air streams: a fast one, 10 cm in height, simulating outdoor wind speeds up to 3 m s^{-1} above plant stands or bare soil surfaces and a slow stream, 80 cm high, simulating air exchange rates in plant stands (up to 0.3 m s^{-1}). As in the application chamber, the experimental platform is variable in height and is fixed before the start of an experiment so that the surface area (bare soil or upper part of a plant stand) reaches into the fast air stream. During experiments with bare soil, the wind channel producing the slow wind speed is closed, so that real outdoor conditions can be simulated for a wide range of applications.

After having passed the chamber room, the two air streams reach mixing channels in which the air, which may now be loaded with volatile compounds, is homogenized. High-volume sampling systems are installed downstream which allow for a homogeneous isokinetic air sampling up to 10% of a total air volume per time unit. After that the mixing channels reach a filter system normally filled with activated carbon holding back volatile radioactive compounds. A second blower discharges the air from the system. The air pressure in the chamber is equal to the actual air pressure outside because a blower pressing air into the air conditioning system and the second blower pulling at the end compensate each other.

Since irradiation may influence the fate of pesticides applied under field conditions, a metal halide lamp producing up to 1200 W m^{-2} on the plant or soil surface is installed above the chamber which is covered with a Sanalux$^{®}$ glass allowing for the transition of UV-light with small losses only. This lamp produces a wavelength distribution very similar to natural sunlight. The lamp is computer-controlled and its light intensity can simulate the solar intensities from sunrise to sunset. Figure 2 illustrates the complete volatilization chamber.

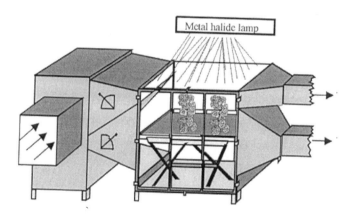

Figure 2. Design of the volatilization chamber.(Reproduced with permission of reference 10. Copyright 1999 Verlag Dr. Köster)

Air Sampling and Analysis

For exact investigation of the volatile radioactivity, the sampling systems downstream the isokinetic samplers consist of the following parts:

1. polyurethane foams for adsorption of non-polar volatile compounds. Extraction efficiency and break-through of the compounds is tested separately before the start of an experiment.
2. The trap for sampling any $^{14}CO_2$ which might have been formed as a result of mineralisation of the ^{14}C-labeling position. This trap is a mixture of ethanolamine, 2-phenylethylamine, diethylengycolmonobutylether and ethylenglycole (1:1:1:1, vol:vol:vol:vol).
3. A freezing trap, collecting the air moisture, which can be investigated quantitatively for any polar compounds by LSC.

Together with the investigation of the soil and/or plant material at the end of an experiment a total balance of the applied pesticide can be determined.

Table I. Active Ingredients used for the Experiments

	Fenpropimorph	Parathion-methyl
IUPAC-name	(±)-cis-4-[3-4(tert-butyl-phenyl)-2-methylpropyl] -2,6-dimethylmorpholin	O,O-dimethyl-O-4-nitrophenyl-phosphorus-thioate
Sum formula	$C_{22}H_{33}NO$	$C_8H_{10}NO_5PS$
Molar mass [g mol^{-1}]	303.5	263.2
Vapor pressure [Pa]	3.5×10^{-3}	1.29×10^{-3}
Water sol. [mg L^{-1}]	4.3	55.0
Henry's law constant	1.0×10^{-7}	2.5×10^{-6}
Log Pow	4.1	3.0
^{14}C-labeling position	[U-^{14}C]benzolring	[U-^{14}C]phenylring
Specific radioactivity [kBq mg^{-1}]	30 - 75	420
Formulation type	EC	WP
A.I amount [g ha^{-1}]	750	200

Pesticides used and experiments carried out

Experiments were carried out using the active ingredients described in Table I. A loamy sand (sand: $\geq 70\%$, clay: $\leq 10\%$, Corg: $\leq 1.5\%$) was used for the experiments. French beans in the 10-20 leaf stage just before blossom and barley just before ear emergence, grown under greenhouse conditions were used for the studies. To distinguish exactly between volatilization from plants and soil in case of plant experiments the soil was covered with filter paper during application. This paper was removed before the start of the experiment and the amount of ^{14}C was determined after extraction and combustion of the filter paper. A survey of the experiments carried out is given in Table II.

Table II. Experiments carried out

AI	No. of replicates	Target	Climate	Duration [h]
Parathion-methyl	2	French beans	Simulated day	24
Parathion-methyl	1	French beans	Simulated day without irradiation	24
Fenpropimorph	3	Soil/barley	Simulated day	96

Sample work-up and analysis procedure

The PU-foams were changed after 1, 3, 6 and 24 h and in case of the experiments with fenpropimorph in addition after 48, 72 and 96 h. The freezing and $^{14}CO_2$-traps were changed every 24 h. After 24 h (parathion-methyl) or after 96 h (fenpropimorph) the experiments were stopped, the plants were harvested and the soil was taken out for homogenization. 100 g samples were taken from the homogenized soil for extraction and ^{14}C-measurement. Extraction of fenpropimorph and metabolites from soil was carried out using chloroform and a Bleidner-apparatus (11) using a method described by Heizler (12). Plants were extracted with methanol. In the case of parathion-methyl extraction was carried out using acetone. For chromatographic characterization radio-HPLC (fenpropimorph) or radio-TLC (parathion-methyl) was used (13, 14). All PU foams were extracted following a procedure described by Niehaus (15) and aliquots of the frozen air humidity and the $^{14}CO_2$-traps were measured using suitable scintillators.

Results

Experiments with Parathion-methyl

Parathion-methyl experiments were carried out for 24 h simulating a measured weather scenario in May in Germany. In two experiments the irradiation influence was taken into account and in one additional experiment it was not. Figure 3 shows the measured and simulated scenarios.

264

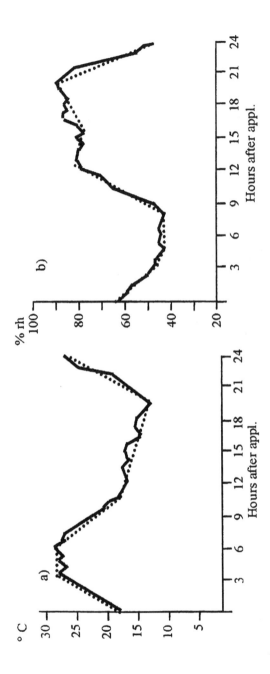

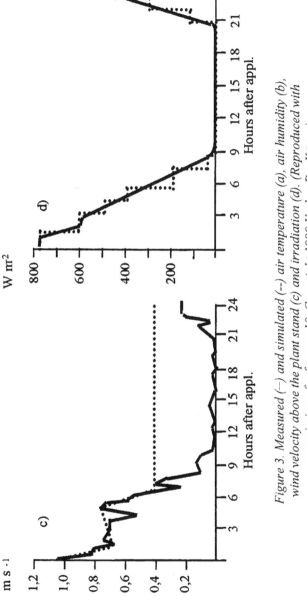

Figure 3. Measured (–) and simulated (–) air temperature (a), air humidity (b), wind velocity above the plant stand (c) and irradiation (d). (Reproduced with permission of reference 10. Copyright 1999 Verlag Dr. Köster)

Table III. Volatile radioactivity after application of ^{14}C-parathion-methyl to French beans.

Experimental Conditions	Volatilization [%][1] after			
	1h	2h	6h	24h
without irradiation	15.9	40.5	53.4	77.2
with irradiation (1st repl.)	15.3	45.6	60.2	65.9
with irradiation (2nd repl.)	19.9	49.9	56.1	65.3

[1]:radioactivity applied = 100 %

Under these conditions 15.3 % to 19.9 % of the ^{14}C-parathion-methyl applied volatilized during the first hour after application. After 3 hours already 40.5 % and 49.9 % had volatilized. These processes continued, and after 24 hours more than 60 % of the radioactivity applied was determined in the PU-foams of the trap system in the 3 experiments (13). Table III shows the volatilization kinetics of both experiments under the influence of simulated sunlight and the kinetics of the volatilization of radioactivity in the dark.

At this time, the 3 experiments were stopped, the plant material was harvested, extracted and analyzed. Table IV provides the mass balances.

Table IV. Mass Balances of Experiments with ^{14}C-Parathion methyl

Kind of experiment	Volatilized after 24 h [%][1]	Extracted from plants [%][1]	Not extracted from plants [%][1]	Mass balance [%][1]
Simulated outdoor conditions including irradiation (1st repl.)	65.9	20.9	8.0	94.8
Simulated outdoor conditions including irradiation (1st repl.)	65.3	19.4	8.4	93.1
Simulated outdoor conditions without irradiation	77.2	17.3	3.0	97.5

[1]: radioactivity applied = 100 %

The radioactivity extracted from plants as well as the amount of ^{14}C volatilized and sorbed on the PU-foams were further investigated by TLC. The results showed that more than 99 % of the volatilized ^{14}C represented still unchanged parathion-methyl in all experiments. Furthermore, more than 99 % of the radioactivity extracted from the French beans was still the A. I. whereas under the influence of simulated sunlight more than 50 % of the radioactivity extracted were polar metabolites so that the amount of parathion-methyl still available on the plants was reduced (10).

Experiments with fenpropimorph

For the fenpropimorph experiments a typical four-day German weather scenario was recorded outside and programmed. Figure 4 shows the measured and simulated data.

Figure 5 characterizes the results of the volatilization kinetics of ^{14}C-fenpropimorph from the plant/soil systems. The volatilization rate was rapid within the initial 24 hours after application in the three experiments. Subsequently volatilization rate decreased and was nearly complete after 4 days.

Table V. Mass Balance of Experiments with ^{14}C-Fenpropimorph

Experiment	I [%][1]	II [%][1]	III [%][1]
Volatile ^{14}C	48.1	46.1	60.3
Volatile $^{14}CO_2$	1.8	1.1	0.0
Extracted from plants	15.8	15.2	14.5
Not extracted from plants	6.5	8.2	3.9
Extracted from soil	12.1	15.3	24.8
Not extracted from soil	6.2	4.0	1.5
Mass balance	90.5	89.9	105.0

[1]: radioactivity applied = 100 %

The mass balance in these experiments consisted of the extractable radioactivity in soil and in plants and the volatile compounds. In addition to the volatile compounds determined in the PU-foams in two of the three experiments small amounts of $^{14}CO_2$ were detected in the CO_2 traps. Table V shows the details of the results obtained.

Most of the ^{14}C-labeled residues recovered were extractable from plant and soil. The results obtained from HPLC investigation showed that at least 60 % of the radioactivity extracted from plants consisted of the metabolite fenpropimorph acid and other polar metabolites (11). In the soil extracts, however, only the unchanged fenpropimorph was determined (14).

Discussion

The application equipment presented here provided for an application procedure similar to agricultural practice. This was shown by determination of deviations of spraying amount per 100 cm^2 to be 9 %, i.e., an equal distribution of the spray volume in this equipment is possible with nozzles licensed for agricultural use (16). In the same paper, the wind profile in the chamber was investi-

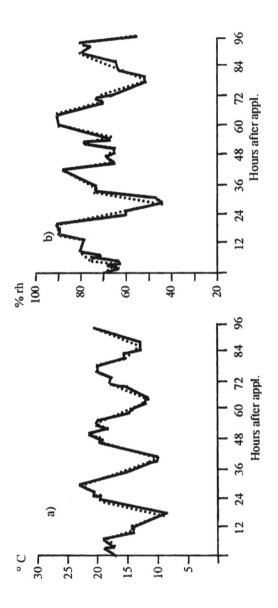

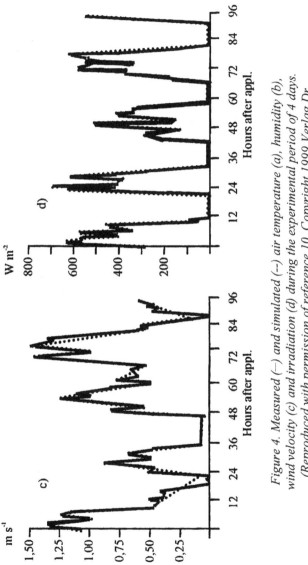

Figure 4. Measured (—) and simulated (--) air temperature (a), humidity (b), wind velocity (c) and irradiation (d) during the experimental period of 4 days. (Reproduced with permission of reference 10. Copyright 1999 Verlag Dr. Köster)

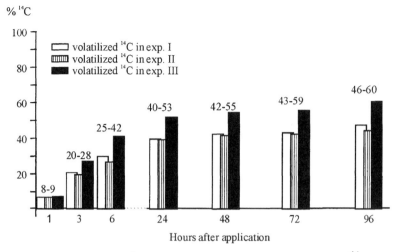

% ^{14}C

Figure 5. Kinetics of volatile radioactivity after application of ^{14}C-fenpropimorph to barley/soil. (Reproduced with permission of reference 10. Copyright 1999 Verlag Dr. Köster)

gated. Therefore, the wind speed was measured at different heights at the chamber front, in the middle and in the back of the chamber, aspiring a speed of 1 m s^{-1} at a height of 80 to 90 cm and an air exchange rate of 0.1 to 0.2 m s^{-1} at the chamber bottom. The results showed, that the wind speeds were uniform at each point and were rapidly reduced in the transition zone (area of plant stand surface). These conditions were also measured in fields with barley and wheat (10).

Since it is well known that results of laboratory experiments do not reflect the real outdoor conditions in detail, (17) the volatilization chamber presented here cannot simulate a field experiment but rather the parameters which have an important influence on the metabolism and volatilization of compounds applied in agriculture. This is the air temperature, the wind velocity, the relative air humidity and the irradiation. It could be shown that the outdoor scenarios recorded and programmed were simulated very well and were reproduced exactly. The only restrictions are that wind velocities near calm can not be simulated because the online air condition system requires a minimum of air exchange, and that the intensity of the metal halide lamp can be adjusted only stepwise (10).

The results obtained with ^{14}C-parathion-methyl and ^{14}C-fenpropimorph showed that volatilization may be a rapid and decisive process immediately after application and may be of importance for the interpretation of residue analysis in field dissipation studies. The experiments with ^{14}C-parathion-methyl indicated that the influence of irradiation may be of importance already after application because photolytical processes of the compounds sorbed on the target can appear. Photolytical reactions of volatilized compounds in the atmosphere can not be simulated in this chamber which is another restriction of the system. This is due to the fact that the time the volatilized chemicals stay in the chamber is too short for photodegradation.

In recent years several attempts were made to measure volatile compounds directly in the atmosphere after their application to the field. Therefore meteorological distribution models (18, 19) are used. It was already shown that these models may be used for a mass balance, however, with the restriction that these experiments require enormous preparations and special environmental conditions, e.g., concerning the wind direction, which are often not available in the field. Therefore, the volatilization chamber described here provides an easy-to-use experimental tool which supplies exact and reproducible information about the further fate of a pesticide after application taking into account mass balance, degradation and transport. In order to investigate whether the results from the volatilization chamber reflect the field conditions, an experiment was started where a wettable powder formulation of parathion-methyl was applied to French beans (20). Directly after application and after 1, 3, 6 and 24 hours randomized plant sampling was carried out and the parathion-methyl concentration was determined. The environmental conditions during the experiment were measured and recorded and used for a parallel experiment with ^{14}C-parathion-methyl in which the application conditions as well as the weather conditions during the experiments were repeated exactly. Volatile ^{14}C-parathion-methyl was measured after PU-foam extraction as described above. The results showed that the lack of plant residues compared with the initial value determined in the field experiment by residue analysis after 1, 3, 6 and 24 hours were in good accordance: After 1 hour 86.1 % of the initial value were determined in the field experiment and 15.9 % volatilization were measured in the corresponding chamber experiment. After 3 hours 48.6 % of the initial value were determined on the plants in the field whereas 41.4 % volatilization were measured in the laboratory. Six hours after application 29.8 % had remained in the field and 53.4 % had volatilized in the chamber. After 24 hours all plants were harvested and analyzed. At this time 25.9 % of the initial value were determined in the field and 77.2 % of the ^{14}C applied had volatilized in the laboratory. Together with the extractable and non-extractable residues in the plants the mass balance of the chamber experiment was 97.5 %.

Experiments in the volatilization chamber can also contribute to a better interpretation of field dissipation studies if not only volatilization occurs but also mineralisation. This was demonstrated in an experiment with ^{14}C-dimethoate after application to soil (20). After an experimental time of 24 hours, 7.9 % $^{14}CO_2$ appeared in the CO_2-traps of the chamber resulting from an intensive biodegradation by soil microorganisms. Another 7.8 % were no longer extractable from soil, 10.5 % were metabolites and only 1.3 % were determined as volatile dimethoate. Together with the extractable dimethoate residues (75.4 %) a mass balance of 102.9 % was obtained.

The results presented here show that the volatilization chamber can reflect the field situation concerning the parameters important for the fate of pesticides. It provides detailed information for a better understanding, the results are reproducible and mass balances can be prepared when using ^{14}C-labeled compounds .

References

1. Gath, B.; Jaeschke, W.; Kubiak, R.; Ricker, R.; Schmider, F.; Zietz, E. *Nachrichtenbl. Deutscher Pflschtzd.* **1993**, *45*, 134-143

2. Wienhold, B. J.; Gish, T. J. *J. Environm. Qual.* **1994**, *23*, 292-298

3. Harper, L. A.; White, A. W.; Bruce, R. R.; Thomas, A. W.; Leonard, A. A. *J. Environm. Qual.* **1976**, *5*, 236-242

4. Guenzi, W. D.; Beard, W. E. *Soil Sci. Soc. Am. Proc.* **1970**, *34*, 443-447

5. Grass, B.; Wenzlawik, B. W.; Rüdel, H. *Chemosphere.* **1994**, *28*, 491-499

6. Hock, B.; Fedtke, C.; Schmidt, R. R. Herbizide: *Entwicklung, Anwendung, Wirkungen, Nebenwirkungen*; Thieme-Verlag: Stuttgart, GE, 1995

7. Van den Berg, F.; Kubiak, R.; Benjey, W. G.; Majewski, M. S.; Yates, S. R.; Reeves, G. L.; Smelt, J. H.; Van der Linden, A. M. A. In *The Fate of Pesticides in the Atmosphere;* Van Dijk, H. F. G., Van Pul, A. J., De Voogt, P.; Kluwer Academic Publishers: Dordrecht, NE, 1999, 195-218

8. Breeze, V. G.; Fowler, A. T. *New Phytol.* **1992**, *120*, 389-396

9. Kubiak, R.; Maurer, T.; Eichhorn, K. W. *Sci. Of Total Environm.*, **1993**, *132*, 115-123

10. Kubiak, R. *Praxisrelevante Untersuchungen zum Verhalten von Pflanzenschutzmitteln in der Umwelt*; Schriftenreihe Agrarwissenschaft Band 6; Verlag Dr. Köster, Berlin, GE, 1999, 1-155

11. Staimer, N.; Müller, T.; Kubiak, R. *Int. J. Environm. Anal. Chem.* **1996**, *65*, 183-191

12. Heizler, W. *Methodensammlung zur Rückstandsanalytik von Pflanzenschutzmitteln*; Deutsche Forschungsgemeinschaft; Verlag Chemie; Weinheim, GE, 1991

13. Müller, T.; Maurer, T.; Kubiak, R. *Environm. Monitoring and Assessment.* **1997**, *44*, 541-547

14. Müller, T.; Staimer, N.; Kubiak, R. *Pestic. Sci.* **1998**, *53*, 245-251

15. Niehaus, R.; Scheulen, B.; Dürbeck, H. W. *Sci. Of Total Environm.* **1990**, *99*, 163-172

16. Maurer, T.; Kubiak, R. *Ecoinforma.* **1994**, *9*, 567-579

17. Kubiak, R.; Führ, F.; Mittelstaedt, W.; Hansper, M.; Steffens, W. *Weed Sci.* **1988**, *36*, 514-518

18. Majewski, M. S.; Desjardins, R.; Rochette, P.; Pattey, E.; Seiber, J. N.; Glotfelty, D. E. *Environm. Sci. Technol.* **1993**, *27*, 121-128

19. Majewski, M. S.; Glotfelty, D. E.; Paw, K. T.; Seiber, J. N. *Environm. Sci. Technol.* **1990**, *24*, 1490-1497

20. Kubiak, R.; Müller, T.; Maurer, T.; Eichhorn, K. W. *Int. J. Environm. Anal. Chem.* **1995**, *58*, 349-358

Chapter 18

Determining the Dissipation and Potential Offsite Movement of Pendimethalin through a Monitoring Survey and Environmental Modeling

Patricia J. Rice, Gary D. Mangels, and Maximilian M. Safarpour

Department of Environmental Chemistry and Metabolism, BASF Agro Research Division, Princeton, NJ 08543–0400

Dissipation, accumulation, and the potential offsite movement of pendimethalin, a soil-applied herbicide, were determined by conducting a retrospective monitoring survey and environmental modeling. The test area for the monitoring survey had a history of annual pendimethalin applications. Soil, water, and sediment samples were collected and analyzed. Information from laboratory and field dissipation studies was utilized in the environmental modeling to predict potential long-range transport and deposition to soil, water, and sediment. The monitoring survey has shown very low concentrations of pendimethalin in the treated soil. Residues were not detected in the non-agricultural soil, surface water or sediment, which indicates no accumulation of pendimethalin, at levels greater than the limit of detection, as a result of degradation, offsite movement through runoff and/or atmospheric transport and re-deposition. Model simulations estimate that the concentration of pendimethalin in soil, water, and sediment resulted in concentrations below the level of ecological concern.

Pendimethalin, a pre-emergence and early post-emergence herbicide used to control broad leaf weeds, has been applied to soil throughout the world for over 25 years. Currently this compound is undergoing reregistration in Europe. In the evaluation process concerns have been raised that long-term use, persistence, and slow volatilization of pendimethalin may result in soil accumulation, off-site movement, and atmospheric transport and deposition. Laboratory studies indicate pendimethalin persists in soil and slow volatilization (3-7% applied dose within 4 months) occurs over time (1-4). European terrestrial field dissipation studies have shown moderate persistence of this compound in soil with half-lives ranging from 27 to 155 days (84 day average) and dissipation resulting from degradation and volatilization rather than leaching (1).

As a result of these concerns, a retrospective monitoring survey of a site, in Spain, with a history of annual application of pendimethalin was performed to evaluate the environmental fate of this herbicide. The objectives of the monitoring survey were to collect and analyze soil, water, and sediment samples from the lower Tietar Valley to determine: 1) soil concentrations and potential accumulation of pendimethalin in treated agricultural soils, and 2) the potential transport, deposition, and accumulation of pendimethalin in non-agricultural areas of the watershed as a result of offsite transport.

For the environmental modeling a Level III fugacity model was utilized to predict the potential long-range atmospheric transport of pendimethalin to remote areas due to volatilization and deposition. Fugacity models have been shown to accurately predict the atmospheric transport of several organic chemicals (5). This chapter summarizes the monitoring survey and atmospheric transport modeling that was conducted for Europe.

Experimental Design

Monitoring Survey

The retrospective monitoring survey was conducted in December 1999 in the Tietar Valley in Talayuela, Spain. The test area had annual applications of pendimethalin to agricultural fields. Soil samples were collected from the treated fields and remote non-agricultural land (several kilometers down wind from treated field) within the lower Tietar Valley. Water and sediment was sampled at various sites along the Tietar River and Torrejon Tietar reservoir, which were located adjacent to, and down stream from, the pendimethalin treated fields. Selection of the test site, site characteristics, agricultural practices, location of sampling sites, and the sample collection and analysis procedures were recorded in Rice et. al. (2000) (6) and are briefly discussed in the following text.

Test Site. The survey was conducted in the Tietar Valley, in Talayuela, Spain. The Tietar Valley contains large areas of intensive cultivation along with nonagricultural land used for pastures and the growing of oak trees. Seasonal brooks run through the treated agricultural land and empty into the Tietar River, which is adjacent to many agricultural fields. An average of 13 tons of pendimethalin are applied in the Talayuela area each year, which accounts for 20% of the annual pendimethalin use in Spain. This area was selected based on several factors which include: 1) highest average annual pendimethalin application in the last five years within a province, 2) largest concentration of pendimethalin used within the sub-regions of the province with the largest load (tons/year), 3) proximity of the water courses through the treated site, 4) significant erosion factors: field slopes and rainfall intensity, 5) high cultivation intensity and occurrence of repeated treatments in the same fields, and 6) potential areas of sedimentation including dams or other water retention points within the water system.

Sample Collection

Soil from Pendimethalin-treated Agricultural Fields

Sample Sites. Soil samples were collected from agricultural fields located within the test area. Fifteen representative agricultural fields with a history of pendimethalin use were selected for the survey. These fields were in crop production and had received annual applications of pendimethalin over the last five years, with the exception of one field that was not treated in 1998. The treated fields were characterized as sand, loamy sand, or loam soil and were planted with tobacco for five consecutive years, with the exception of one field that was rotated with peppers. Three types of applications were distinguished: pre-transplanting, between rows, and de-suckering (Table I). Annual applications of pendimethalin ranged from 0.83 to 2.31 kg/ha (6).

Sample Procedure. Soil samples were collected in December 1999, approximately 4 to 8 months after the last pendimethalin application. Ten soil cores were taken at random within each field from undisturbed soil using 30 cm x 23 mm diameter PETG zero contamination tubes (model JMC Backshaver Handle). The upper and lower ends of the cores were clearly marked. Each soil core was labeled, and placed in freezer within hours of sampling. Samples remained frozen during shipping and storage, prior to analysis.

Table I. The application history of pendimethalin to fifteen agricultural fields (1995-1999)

Type of Application	Rate (kg/ha)[a]	Time of Application	Application Technique
Pre-transplanting	0.83 - 1.32	April/May	Broadcast Spray/Incorporation[b]
Between Rows	0.5	June	Directed Spray
De-suckering	0.5 – 0.99	August	Broadcast Spray

[a]Range of annual pendimethalin application rates within the fifteen fields
[b]Pendimethalin was sprayed onto the soil surface and incorporated to a depth of 3-7 cm

Soil from Untreated Test Area

Sampling Sites. Soil samples were collected from remote, nonagricultural areas located within the test site. Five representative areas with no history of pendimethalin use were selected for the survey. The nonagricultural areas were pastures with grasses, leguminous plants, rockroses, thyme and a few hoalm oak trees. Soil textures ranged from loamy sand to clay (6).

Sample Procedures. Soil within the untreated test area was compact since this non-agricultural area was not regularly tilled. Therefore, bulb planters were used to collect the soil rather than the soil core samplers mentioned above. Ten 5-cm deep soil samples were randomly collected from each untreated test area using bulb planters. Different bulb planters were used at every site to prevent cross-contamination of the sample media. Ten soil samples from each site were combined to make one representative sample. Soil samples were stored frozen until analysis.

Water and Sediment Samples

Sampling Sites. Water and sediment samples were collected at 14 locations from the Tietar River and reservoir system within the test site. Control samples were collected from one location in the mountain area, Sierra de la Serrana, which is a natural reservation without agricultural use. The accumulation of sediment particles, that may be a result of agricultural runoff, were found at regions of the river where water flow was slow. These regions were located in areas where the river branched, along the outer bank at the bend of the river, and near rock formations or small islands.

Sampling Procedures. Water and sediment samples were taken from each sampling site. The sampling technique depended on the depth of the river and the nature of the riverbed. Water samples were collected prior to sediment in order to minimize the quantity of sediment particles in the water layer. Immediately after sampling, several physical parameters (pH, temperature, Redox, conductivity, dissolved oxygen) were measured to characterize the nature of the sample material. Detailed descriptions of the sample locations and observations have been documented (6).

Collection of Water Samples. Depending on the depth of the water, the water samples were collected either by means of a sampling device (System Dr. Blasy) or by filling the glass bottles directly. Each water sample consisted of approximately 1200 ml of water. At least two replicate water samples were collected at every sampling site.

The Dr. Blasy water-sampling device was used from a boat or a bridge. Before taking the water sample the depth of the river was measured using a plumb line. The Dr. Blasy device consisted of a sampling tube, an upper seal, lower seal with a drain, clip, and a messenger that were suspended from a cable. The decontaminated sampling device was preset with the upper and lower seals pulled away from the sampling tube, allowing the water to pass through the tube while it was lowered to the predetermined depth (mid-depth of water body). After the sampler had reached the required depth, the messenger was sent down to release the tube and to enclose the water sample between the seals. Water samples were retrieved and stored in labeled glass bottles. The procedure was repeated until sufficient sample material was collected.

The direct sampling technique was utilized in shallow areas of the river (< 3 meters deep). Sample bottles were submerged in the water with the opening held down. Once the bottle was mid-depth in the water, it was turned 180 degrees so that the opening faced the surface of the water and water could flow into the bottle.

Collection of Sediment Samples. The sampling technique for the sediment samples depended on the depth of the water, and the nature of the sediment and the riverbed at the corresponding position.

The Ekmann-Birger dredge consisting of a sample container to entrap sediment in spring-loaded jaws was arranged so that the jaws were in open position. The sampler was lowered to a point just above the sediment surface by means of a cable and dropped sharply onto the sediment. The release mechanism was triggered by lowering the messenger down the line. The sampler dredge was raised and the sediment was transferred into a stainless steel bucket where it sat undisturbed until the sediment particulates had settled out of the water. The supernatant water was carefully decanted and the required amount of sediment was transferred into two labeled polyethylene containers. These containers were considered the two replicate sediment samples from each sampling site.

A stainless steel bucket was used to collect sediment from beneath a shallow water layer. The decontaminated bucket was used to scoop the top layer of sediment from a designated area. The sediment sample was collected in one continuous motion starting down stream and moving up stream to prevent the loss of any fine sediment particulates. The sediment was handled and placed into the appropriate containers as stated above.

Sample Preparation and Analysis

Soil, water, and sediment samples were stored frozen until analysis (< 1 month after collection). The sample preparation and analysis are summarized below. Samples were extracted and the extracts were analyzed on a gas chromatograph (GC) equipped with an electron capture detector (ECD). The presence of pendimethalin residues was confirmed by GC/MS/MS. BASF (formerly American Cyanamid Company) supplied the pendimethalin analytical standard.

Soil. The top 5-cm of the ten cores from each treated field were removed, combined, homogenized, and placed in labeled containers. These soil samples were combined and sieved through a 3.5 mm mesh sieve to remove stones and debris and were thoroughly mixed to further homogenize the soil. Aliquots of the soil samples were analyzed for pendimethalin according to validated BASF (formerly American Cyanamid) Method M 2514 (limit of quantification (LOQ) = 10 ppb, limit of detection (LOD) = 2 ppb) *(7)*. Soil aliquots were extracted with acidic methanol (2%) and filtered. Hydrolchoric acid (0.1 N) was added to aliquots of the filtrate in a 1:1 ratio. The acidified filtrate was then passed through a conditioned C18 solid phase extraction cartridge (SPE). SPE cartridges were washed with deionized water and the pendimethalin was eluted with 1% methanol in hexane. Eluates were evaporated to dryness, redissolved in hexane, and analyzed by GC. Additional soil aliquots were removed and used to determine the moisture content. The same procedure was followed for the 5-20 cm soil core segments. The 20-30 cm portion of the soil cores remained intact and stored in a freezer. Identical protocols were followed to analyze the untreated soil samples, except these samples only consisted of the top 0-5 cm of soil collected with the bulb planters *(6)*.

Water. Replicate water samples from each sampling site were combined and homogenized. Water samples were analyzed according to the validated BASF (formerly American Cyanamid) Method M 1715 (LOQ 0.1 ppb) *(6,8)*. Aliquots of the water samples were passes through conditioned C18 SPE cartridges. These SPE cartridges were washed with deionized water and eluted with hexane. Eluates were concentrated and analyzed for pendimethalin by GC.

Sediment. Sediment samples were allowed to thaw and settle. The standing water at the top of each sample was carefully removed and the sediment was passed through a 3.5 mm mesh sieve to remove stones and debris. The sieved sediment was thoroughly mixed. Subsamples of each sediment were removed and analyzed for pendimethalin according to BASF (formerly American Cyanamid) Method M 2514 (6,7) as previously described for the soil samples.

Modeling Long-Range Atmospheric Transport and Deposition

A Level III fugacity model was used to evaluate the potential long-range transport of pendimethalin to remote areas due to volatilization and deposition (1). We selected this model since it had been shown to accurately predict the atmospheric transport of several organic chemicals (5). Mackay et al. (5,9-11) described the Level III fugacity model and its usefulness to evaluate the environmental fate of both new and existing chemicals. Each model simulation represented a 100,000-km^2 region (approximate area of Greece), which was composed of four major compartments: soil, air, water, and sediment. The soil consisted of the top 10 cm of soil with a 2% organic carbon content. The air was considered the 1000-m of atmosphere directly above the region. Water accounted for 10% of the surface area of the region and had a depth of 20 m. The sediment had a depth of 1 cm and a 4% organic carbon content (5,9)

The average annual application of pendimethalin in Europe is 127,000 kg/100,000 km^2 of area (1). Because the fugacity model is a steady state model, the mass of pendimethalin input into the system needs to be a constant value over time. Since the average field dissipation half-life in European soil is approximately 84 days and pendimethalin can be applied in both the spring and fall, it is reasonable for an initial assessment to simulate the applications of pendimethalin to the soil as occurring uniformly over time. Therefore pendimethalin was simulated as being introduced into the soil at a rate of 40 kg/hour, that corresponds to a total loading of 350,400 kg/100,000 km^2 and represents the highest loading in any major region in Europe. This total loading represents a realistic "worst-case" scenario for the use of pendimethalin. Residues of pendimethalin remain in the top centimeters of the soil profile, since pendimethalin is strongly adsorbed by soil. Most applications in Europe are made to the surface of the soil and are not mechanically incorporated. Therefore, the environment that was modeled was changed so the depth of the soil was 2.5 cm rather than the 30-cm depth routinely used in the model. This change was performed in order to ensure that potential movement of pendimethalin from the soil into the atmosphere was not underestimated by simulating incorporation into the soil profile. Results from this revised modeling scenario can therefore be expected to reflect or overestimate a real-world scenario.

A series of modeling simulations were run with a combination of pendimethalin half-lives in various phases and the properties that are listed below. Atmospheric half-lives of 4 and 28 hours, which correspond to the rate of reaction with OH radicals as calculated by the method of Atkinson and the rate of direct photolysis corrected for sunlight intensity in Europe, were evaluated (12,13). Simulations included water half-lives of 7 and 28 days, which bracket the rates of degradation due to photolysis and biotic degradation in water (1). The DT50 range of 84 (2016 hours) to 365 days (8760 hours) was selected to represent the average half-life determined from the European field dissipation studies and a conservative estimate of the longest DT50 (331 days) calculated in a laboratory aerobic soil metabolism study (2). A sediment half-life of 14 days (336 hours), determined by a water/sediment study, was used for all simulations (14). This accelerated degradation rate observed in the sediment is a result of the rapid degradation of pendimethalin under anaerobic conditions (14,15). The properties of pendimethalin used in the evaluation were as follows: molecular weight of 281.3 g/mol, melting point of 47 °C, Log K_{ow} of 5.18, water solubility of 0.275 ppm (mg/l), and a vapor pressure of 0.00125 Pa at 25 °C (1,15,16).

Results and Discussion

Monitoring Survey

Pendimethalin-treated Agricultural Fields. Pendimethalin residues were detected in the soil samples from fields containing a history of annual pendimethalin applications. Soil residues in the 0-5 cm and 5-20 cm depths are listed in Table II. Concentration of pendimethalin within the top 5 cm ranged from < 10 ppb (LOQ) to 194 ppb. The 5-20 cm soil contained pendimethalin residues between < 10 ppb (LOQ) to 73 ppb with 40% of the analyzed samples below the limit of quantification. If no degradation/dissipation of pendimethalin occurred in the soil following an annual application and the total quantity applied reached the soil surface, the estimated concentrations of 1107 and 3080 ppb (µg/kg) would be expected within the top 5 cm of soil. Pendimethalin has a low water solubility (0.275 ppm) and is strongly sorbed to soil (Koc > 13,000 ml/g), therefore its tendency to leach is minimal (15). Cultivation practices within these fields (disc harrow, 15-20 cm deep, autumn and early spring) can result in an incorporation of pendimethalin down to a depth of 15 to 20 cm (6). The pendimethalin detected in the 5-20 cm segments are believed to be a result of the agricultural practices rather than downward movement with water.

Table II. Pendimethalin Residue Detected in the Treated Field Soil
(0-5 cm and 5-20 cm depths)

Treated Field No.	Residues at 0-5 cm depth (μg/kg)[a]	Residues at 5-20 cm depth (μg/kg)[a]
1	194[b]	25[b]
3	67[c]	33[c]
4	30[b]	15[b]
5	175[b]	26[b]
8	55[c]	16[c]
9	<10[d]	<10[d]
10	13[b]	<10[b]
13	112[c]	<10[c]
14	13[c]	<10[c]
15	148[c]	73[c]

[a]Concentration based on soil dry weight (μg/kg = ppb)
[b]Last application was applied 4 months prior to sample collection
[c]Last application was applied 7-8 months prior to sample collection
[d]Last application was applied 6 months prior to sample collection

To further determine if pendimethalin accumulates in the soil after repeated annual applications, the total residues detected in the soil (total ppm within 0-20 cm depth) were compared with the total quantity of pendimethalin applied to the soil during the past five years (1995-1999) Table III. Less than 0.1% of the pendimethalin applied to the agricultural fields, within the last five years, remained in the soil. These results indicate pendimethalin is degrading and/or dissipating from the soil, and is not accumulating as a result of the repeated annual applications. Pendimethalin may also adsorb to the soil and therefore be less bioavailable.

Untreated Test Area. Analysis of water, sediment, and untreated soil samples show no detectable residues of pendimethalin (water < 0.1 ppb, sediment and soil < 10 ppb). Recoveries of the spiked water (0.05 ppb), sediment (10 ppb), and soil (10 ppb) samples ranged from 89% to 103% (1). The data indicates no accumulation (> LOD) of pendimethalin in the water/sediment of the Tietar River/Reservoir and soil from nearby untreated fields as a result of off-site movement, transport, and deposition. Pendimethalin has a low water solubility (0.275 ppm) and readily adsorbs (Koc > 13,000 ml/g) to soil (1,15). As a result, minimal off-site movement to adjacent surface-water systems should occur unless the herbicide is carried by soil particles after a heavy rain. In addition, pendimethalin has been shown to readily degrade in surface waters as a result of photolysis (15). Therefore, accumulation of pendimethalin residues in surface waters within the treated agricultural area is not to be expected

Table III. Comparison of pendimethalin residues detected in the soil to the total quantity of pendimethalin applied during the last five years (1995-1999)

Field number	Pendimethalin residues (mg/kg)[a]	Total applied in 5 years (mg/kg)[b]	% remaining of 5 year applications
1	0.219	275	0.08
3	0.1	275	0.04
4	0.045	176	0.03
5	0.201	220	0.09
8	0.071	385	0.02
9	ND[c]	319	0.00
10	0.013	286	0.00
13	0.112	165	0.07
14	0.013	308	0.00
15	0.221	281	0.08

[a]Total pendimethalin residues detected in the soil samples (0-20 cm)
[b]Total quantity of pendimethalin remaining in soil if no degradation or offsite transportation occurred during the last 5 years
[c]ND = not detected

Level III Fugacity Model Simulations

The Level III fugacity model has been used to describe the environmental fate of a wide variety of chemicals, including persistent organic compounds (5,10,11). This model provides a reasonable estimate of measured concentrations as can be seen in Mackay and Paterson (1991) (5).

Modeling results were based on the properties of pendimethalin and various half-lives of pendimethalin in soil, water, sediment, and air (Table IV). Soil concentrations were calculated based on a soil depth of 2.5 cm as discussed in the methods. Assuming there is no degradation in soil, the concentration of pendimethalin in the top 2.5 cm of the soil, within the 100,000-km^2 region, is calculated to be approximately 4400 ng/g (ppb). In the field dissipation studies conducted in Europe, soils were collected and analyzed to a depth of 5.0 cm rather than 2.5 cm, therefore the predicted pendimethalin residues would be 2200 ng/g if no degradation had occurred. Modeling simulations, conducted with a soil degradation half-life of one year, predicted concentrations of 145 and 73 ng/g in the top 2.5 and 5 cm of soil. These values were above the limit of detection (LOD = 50 ng/g, for field studies) for soil. Simulations with soil degradation/dissipation half-lives of 180 and 84 days resulted in soil concentrations of 73 and 41 ng/g (ppb) in the top 2.5 cm of soil and 37 and 21 ng/g in the top 5 cm, which were below the LOD for the field studies *(1)*. Therefore, based on results from the modeling, we can conclude that the rate of degradation of pendimethalin in soil under field conditions would be less than one year since predicted soil residues were below the limit of detection used in the field dissipation studies.

Level III fugacity model simulations estimated the concentration of pendimethalin in water and sediment would range from 0.274 - 2.05 ng/l and 0.0174 - 0.13 ng/g (ppb), respectively, depending upon the half-lives in soil, air, and water. Less than 1% of the applied herbicide was predicted to be released into the atmosphere from treated region and less than 0.04% would be deposited onto the soil within the region *(1)*.

Table IV. Environmental Concentrations Predicted by the Level III Fugacity Model

Input				*Results*			
Half-lives (hours)				*Concentrations in*			
Soil	*Air*	*Water*	*Sediment*	*Soil (ng/g)*	*Air (ng/m³)*	*Water (ng/L)*	*Sediment (ng/g)*
8760	28	672	336	145	0.087	2.05	0.13
8760	28	168	336	145	0.0803	0.979	0.0621
8760	4	672	336	145	0.0176	2.04	0.13
8760	4	168	336	145	0.0163	0.975	0.0619
4320	28	672	336	72.7	0.0436	1.03	0.0653
4320	28	168	336	72.7	0.0402	0.491	0.0312
4320	4	672	336	72.7	0.00884	1.02	0.065
4320	4	168	336	72.7	0.00816	0.489	0.031
2016	28	672	336	40.7	0.0244	0.0576	0.0365
2016	28	168	336	40.7	0.0225	0.0275	0.0147
2016	4	672	336	40.7	0.00495	0.573	0.0364
2016	4	168	336	40.7	0.00457	0.274	0.0174

Pendimethalin has been used in Europe for 25 years, allowing sufficient time for the equilibrium conditions that were modeled with the fugacity model to occur. Therefore, the predicted concentrations represent an upper limit. Results of the Level III fugacity modeling have shown that concentrations of pendimethalin in soil, water, and sediment would be minimal and at levels that are below ecological concern for pendimethalin (1).

Atmospheric Transport and Deposition. The amount of pendimethalin moving offsite from a region of 100,000 km^2 into an adjacent region, through transport in the air and deposition, can be estimated by assuming the air over the treated region moves into the adjoining region. Residues of pendimethalin in the air (0.087 ng/m^3) are then subject to degradation in the atmosphere (direct and indirect photolysis) and deposition into soil, water, and sediment. Figure 1 depicts the fate of pendimethalin in the adjacent region, as a result of atmospheric transport and deposition, and represents the "worst case" scenario based on half-lives of 1 year, 28 days, 14 days, and 28 hours in soil, water, sediment, and air, respectively.

This simulation (Figure 1) shows that approximately 94% of the pendimethalin that moved into the adjacent region would be deposited onto the soil resulting in 0.0149 ng/g (ppb) within the top 2.5 cm of soil. An estimated 1.2 and 0.24% of the pendimethalin transported in the atmosphere would be deposited in the water (0.003 ng/L) and sediment (0.0002 ng/g). These levels of pendimethalin are extremely low and are below the levels of ecotoxicological concern for pendimethalin.

Conclusions

The long-term use of pendimethalin provided a situation in which a retrospective monitoring survey and model simulations could be utilized to evaluate the soil dissipation, accumulation, and offsite movement of this widely used herbicide. Environmental concentrations of pendimethalin, in treated soil and adjacent non-agricultural soil, surface water and sediment, were determined after many years of applications under normal agricultural practices. Treated soil contained <1% of the pendimethalin applied to soil during the preceding 5 years indicating degradation and/or dissipation. Pendimethalin may also adsorb to the soil and therefore be less bioavailable. Residues were not detected in the non-agricultural soil, surface water or sediment, which indicates no accumulation of pendimethalin as a result of offsite movement through runoff or atmospheric transport and deposition at the monitored region. In addition, the Level III fugacity model was used to predict potential long-range transport and deposition of pendimethalin from treated soil into adjacent regions. Model simulations

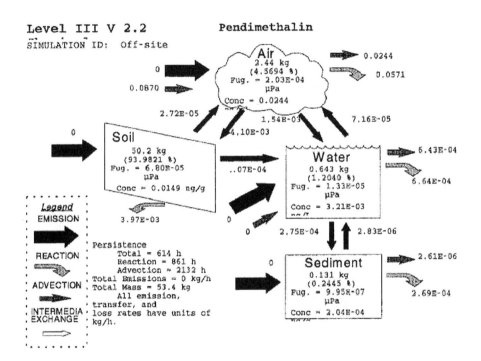

Figure 1. Level III fugacity model simulation of the off-site movement of pendimethalin from soil.

predicted less than 1% of the applied herbicide would be present in the atmosphere at one time and only trace levels of pendimethalin were estimated to deposit into adjacent soil, water, and sediment. Although the results from the retrospective monitoring survey and the atmospheric transport modeling could not be directly compared, due to the difference in size of the test site and modeled region, a congruent conclusion can be drawn: concentrations of pendimethalin in soil, water, and sediment occur at trace levels below ecological concern.

References

1. Mangels, G. BASF confidential report 1, 2000.
2. Yan, Z. BASF confidential report 2, 2000.
3. Voget, M. BASF (formerly American Cyanamid Company) confidential report 3, 1994.
4. Schroll, R.; Dörfler, U.; Scheunert, I. *Chemosphere* **1999**, *39*, 595-602.
5. Mackay, D.; Paterson, S. *Environ. Sci. Technol.* **1991**, *25*, 427-436.
6. Rice, P. BASF confidential report 4, 2000.
7. Singh, S. BASF (formerly American Cyanamid Company) confidential report 5, 1996.
8. Tondreau, R.; Chukwudebe, A. BASF (formerly American Cyanamid Company) confidential report 6, 1987.
9. Mackay, D.; Paterson, S.; Shiu, W. Y. *Chemosphere* **1992**, *24*, 695-717.
10. Mackay, D.; Di Guardo, A; Paterson, S.; Cowan C. E. *Environ. Toxicol. Chem.* **1996**, *15*, 1627-1637.
11. Mackay, D.; Paterson, S.; Cheung, B; Neely, W. B. *Chemosphere* **1985**, *14*, 335-374.
12. Mangels, G. BASF confidential report 7, 2000.
13. Shepler, K.; Mangels, G. BASF (formerly American Cyanamid Company) confidential report 8, 1999.
14. Rudel, H. BASF (formerly American Cyanamid Company) confidential report 8, 1992.
15. *Herbicide Handbook*; Ahrens, W. H., Ed.; Weed Science Society of America Seventh Edition; Weed Science Society of America: Champaign, IL, 1994; pp 230-232.
16. *The Pesticide Manual*; Tomlin, C., Ed.; Tenth Edition; The British Crop Protection Council; Surrey, United Kingdom, 1994; pp 779-780.

Chapter 19

Relating Field Dissipation and Laboratory Studies through Modeling: Chlorothalonil Dissipation after Multiple Applications in Peanuts

R. Don Wauchope[1], Thomas L. Potter[1], and Albert K. Culbreath[2]

[1]Southeast Watershed Research Laboratory, Agricultural Research Service, U.S. Department of Agriculture, Tifton, GA 31794
[2]Department of Plant Pathology, University of Georgia Coastal Plain Experimental Station, Tifton, GA 31794

A computer simulation model can provide useful environmental risk information for pesticides based on rather limited field and laboratory data, if that data provides calibration of the model for the critical processes controlling dissipation. We used the USDA-Agricultural Research Service Root Zone Water Quality Model (RZWQM) to estimate leachate and runoff of chlorothalonil fungicide and its soil degradates during a growing season in which a total of seven applications were made at intervals of 14 days to peanut plants. RZWQM provides detailed algorithms to describe broadcast pesticide application to a combined foliage/soil target, dissipation within both targets, and movement in runoff and leachate water as a function of weather and soil moisture. The model provides an integrated analysis showing how initial partitioning of chlorothalonil between foliage and soil, washoff from foliage to soil, and degradation on the foliage and soil surface (actually the top cm of soil) limit the potential of the parent chlorothalonil and its degradates to leach. However, the model indicates that under severe rainfall conditions during the application period, significant quantities of the parent compound may be transported by runoff into water sources.

This symposium was intended to help define (a) how best to do field dissipation studies, (b) what they actually tell us, and (c) how best to make use of what we already know in order to refine them and minimize the need for them in the future. In most field situations a few processes dominate the dissipation of a majority of the pesticide. If we know what they are and can adequately characterize those processes, we can do a credible job of determining, at least to a first approximation, the probable fate and transport of the pesticide.

Defining the critical process(es) depends on modeling. More than just a description of causes and effects, a realistic pesticide environmental fate model is a *summary of what we know* about the combined effects of multiple interdependent processes occurring in time and space. If our model is even approximately right, it should provide us with strong advice as to the critical measurements that are needed in laboratory and field (1-4). Even more important, an initial modeling exercise with limited data may provide leads as to what the next experiments ought to be. Such an exercise may also serve to validata and calibrate the model. Almost always, application of a model to a new real situation highlights previously unrecognized unrealistic or simplified descriptions of processes in the model.

To illustrate this we describe an experiment in which some exploratory field and laboratory measurements were done on chlorothalonil fungicide (tetrachloro-isophthalonitrile) dissipation in a peanut field soil. These are admittedly only the most preliminary beginnings of a full "field dissipation study". These data were then used to explore the validity of the foliar processes of the ARS RZWQM model(5,6,7), and to define what had been learned in the experiment, and what future measurements need to be conducted in order to adequately describe the fate and behavior of this pesticide in this environment.

Chlorothalonil is a surface protectant fungicide applied to the foliage of a wide variety of crops including peanuts, potatoes, tomatoes and other vegetables, tree-fruits and turf. The product is used widely throughout the world. In the US, peanut production is heavily dependent on chlorothalonil treatments to control several economically significant diseases (8,9). Annual chlorothalonil applications to US peanuts exceed 2.3×10^6 kg, making the crop the nation's number one use of chlorothalonil (8). In the Coastal Plain area of Georgia,

Alabama, and Florida, which is by far the nation's largest-producing peanut region, intense thunderstorms occur during the growing season which generate surface water runoff even though the soils tend to be sandy (10). If this occurs soon after a chlorothalonil application--which is likely, since multiple applications throughout the growing season are typical--we may expect that transport away from the site of application may occur via runoff and/or leaching.

Chlorothalonil Dissipation Experiments

The field and laboratory work are described in detail in (11). Briefly, peanuts were grown in Tift County, Georgia on two-row x 7.6 m plots on a Tifton sandy loam (12). The surface soil is 75% sand, 16% silt and 9% clay with an organic matter content of 0.6%. Peanuts were planted on May 25, 1999 and "inverted" for harvest on October 1, 1999. Irrigation, fertilization, and other pesticide use followed recommend practices. Seven chlorothalonil applications of 1.25 kg/ha each were made at 14-day intervals starting on June 23 at plant emergence and ending on September 16 when plants were in senescence.

Within 4 hours after each chlorothalonil application (with one exception), soil samples were collected from the surface 2 cm of soil and a subsample was analyzed immediately for chlorothalonil and its' degradate compounds. In a few cases soils were sampled after more time had elapsed after application. The remainder of the soil samples was incubated in closed bottles in the laboratory at 30°C and field-capacity moisture content (16% v/v), and sampled for analysis after 1, 4, 7, 14, 21, and 28 days of incubation. Soils were extracted 3X with acetone and the analyte concentrated by reduction of the acetone followed by watering-out the solutes into pH 2 water and then sorption on an OASIS HLB solid-phase extraction cartridge (Waters Co., Milford, MA). Cartridges were eluted with 2 % acetic acid in methanol and acetonitrile, the combined eluents concentrated under nitrogen and treated with 20 % acetic acid in water and fortified with 2-chloroepidine as an internal standard.

Extracts were analyzed by HPLC-MS using an atmospheric pressure chemical Ionization interface and an ion trap mass spectrometer (Thermoquest LCQ DECA system, Thermoquest-Finnegan, San Jose, CA). This procedure allowed the analysis of parent chlorothalonil, its principle soil degradate 4-hydroxy-chlorothalonil, and a series of smaller-yield degradates which for our purposes will be lumped together. Detailed spectral interpretations will be published elsewhere. Chlorothalonil and 4-hydroxy-chlorothalonil recoveries by this method were tested by fortifying blank soil with 100 ppb of both compounds. The mean recovery of the compounds was 86 ± 9 and 94 ± 12 % respectively.

The sorption constant for chlorothalonil and 4-hydroxy chlorothalonil in the Tifton soil was measured in a batch equilibrium experiment. Slurries consisting of 5 g of soil in 25 ml of 0.1 to 0.7 mg/l chlorothalonil or 4-hydroxychlorothalonil in 0.01 M CaCl$_2$ were equilibrated with slow mixing for 2 hours. The short equilibrium time was used because of the rapid breakdown of the parent and the results were used only to provide a rough estimate of soil sorption.

Modelling

Analytical data for the two main species found in the incubated soil samples were visually fit to a first-order degradation model (Figure 1) using finite-difference approximations and the plotting capabilities of a spreadsheet. In this model, parent chlorothalonil (C) is converted to the principle soil degradate 4-hydroxychlorothalonil (COH), which is then converted to other degradates X. Simultaneously, both the parent and hydroxy-degradate are sequestered in soil sites where they form less reactive , but still extractable complexes with soil:

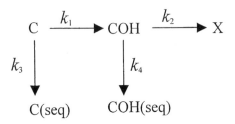

Figure 1. First-order degradation model for chlorothalonil in the top 2 cm of soil. C=chlorothalonil; COH=4-hydroxychlorothalonil; X=sum of degradation products of COH; C(seq),COH(seq) are sequestered C and COH, respectively, which are extractable but not degraded by soil microorganisms

To model the deposition, transport and degradation of these three species within the plant/soil sytem the USDA-ARS Root Zone Water Quality Model (RZWQM) was used. This model includes a detailed functional description of the fate of pesticides, not only in the the "root zone" but at the soil surface and on crop foliage (5,6,7,13).

If the user specifies a broadcast application to soil which is partially covered by crop foliage, the model calculates the interception of pesticide spray by the foliage using the canopy light transmission fraction F_f^{lt} (calculated by the crop growth module of the model). For an application rate of A kg/ha, with an application efficiency factor (fraction of application rate actually depositing on foliage and soil) of F_{ae}, then the resulting total pesticide mass per unit area in $\mu g/cm^2$ is $10 \cdot A$, and the soil and foliar portions are, respectively,

$$C_s^A = 10 \cdot A \cdot F_{ae} \cdot F_f^{lt} \tag{1}$$

and $$C_f^A = 10 \cdot A \cdot F_{ae} \cdot (1 - F_f^{lt}). \tag{2}$$

The pesticide that deposits on the soil surface is assumed to be evenly mixed into the top cm of soil, giving a concentration per unit dry weight ($\mu g/g$) equal to $(C_s^A)/\rho_b$ where ρ_b is the dry soil bulk density at the surface.

Foliar and soil pesticide deposits are tracked separately and one or more pesticide dissipation or degradation processes may be specified for each of the two compartments, or a "lumped" degradation half-life may be specified for each compartment based on field or laboratory monitoring data. Two "daughter" degradation products or one "daughter" and one "granddaughter" product may also be tracked by specifying the parent compound, the compartment(s) and processes which produce them, and the yield fractions, if only part of the degradation of a parent results in daughter compound.

Pesticides and degradates are transferred from foliage to soil by washoff, and within the soil profile by moisture movement (which may be up or down). Washoff is based on the work of Willis et al. (14), who showed that washoff followed an exponential law dependent on the depth of rainfall and the fraction of the deposit available for washoff:

$$C_f = C_f^o \cdot F_f^{wo} \cdot e^{-P_f^{wo} \bullet I_R \Delta t} \tag{3}$$

where C_f and C_f^o are foliar concentrations at time t and $t + \Delta t$ and I_R is rainfall intensity in depth/unit time; F_f^{wo} is the fraction of C_f available for washoff, and P_f^{wo} is an empirical constant that appears to be roughly a function of solubility. Soil water movement is based on the Richards' equation, and pesticide adsorption may be described using either a linear or Freundlich isotherm.

Results and Discussion

Chlorothalonil and 4-Hydroxychlorothalonil Dynamics in Near-Surface Soil

Measured soil sorption coefficients ranged from 5 to 7 for chlorothalonil and 2 for the 4-hydroxide, and we assigned soil organic carbon sorption coefficients of 1600 and 560, respectively, in RZWQM. The chlorothalonil value is reasonably close to the commonly used value of 1380 (15), and the 4-hydroxy-chlorothalonil agrees well with the K_{om} of 250-270 reported by van der Pas, et al. (16), if converted by multiplication by 1.72 to give 430-464, respectively .

The laboratory analytical data for the incubated soils provides a systematic evaluation of how chlorothalonil and degradates accumulate and degrade near the soil surface under peanuts. But the data are complex: multiple applications, foliar washoff and significant metabolite formation will all complicate the picture. The effective sampling depth was 0-2 cm, corresponding approximately to the chemical entrainment "interaction zone" observed in runoff studies (17,18). Analysis (11) indicated that chlorothalonil conversion to 4-hydroxy-chlorothalonil was nearly quantitative, and the 4-hydroxy is then degraded to other compounds. The extractions of later-season soil samples indicated that chlorothalonil applied later in the season was added to residues of both parent and hydroxychlorothalonil which were present from previous applications and which were recoverable by extraction, but were not as readily degraded as the freshly-added chlorothalonil.

The degradation/sequestration model in Figure 1 fits most, but not all of the laboratory incubation data (11). Two examples of data and fitted model curves are shown in Figure 2. Fitted half-lives for chlorothalonil under the warm, moist conditions of the incubations ranged from ranged from 2-4 days in the first two samples to 0.5 - 1.0 day in later samples. It is likely that the first samples, taken from the field when there was no canopy cover, were rather less active microbiologically because the soil had been partially sterilized by the sun. Later samples, taken in the shade under peanut canopy gave more rapid chlorothalonil breakdown. The 4-hydroxychlorothalonil gave a rather consistent 10 to 22 day half life in all samples, with the first two samples again showing less rapid degradation. The half-life for Sequestration of the parent and 4-hydroxy-chlorothalonil were 400 - 2000 and 10-50 days, respectively, though these are very rough estimates used to explain small residuals amounts in most cases.

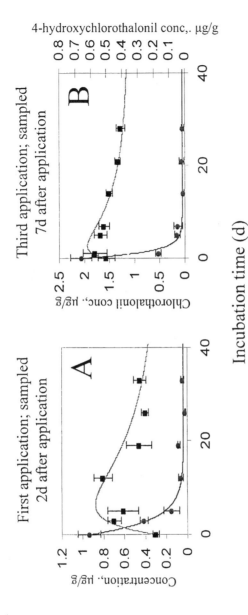

Figure 2. Observed (points) and visual-fit model (curves) for chlorothalonil (circles) and 4-hydroxy-chlorothalonil (squares) residues in 2 soil samples during incubation. (A) First application to soil: model has half lives of 2.3 and 15 days for parent and 4-hydroxy derivative, respectively. (B) A later application which was sampled a week after application; model has half lives of 1 and 10 days for parent and 4-hydroxy compound and "unreactive" sequestered residues of 0.01 and 0.9 ppm, respectively.

Based on these analyses we assigned near-surface, aerobic soil half lives of 2d and 10d to chlorothalonil and 4-hydroxychlorothalonil, respectively, in RZWQM. The former is a considerably shorter half-life than generally reported from field studies of chlorothalonil (15), and must be due to the favorable conditions of the incubation. These conditions are realistic, however, for near-surface soil in this climate. In the case of the 4-hydroxy compound, we were not able to distinguish between degradation and soil binding; see Potter, et al. (11) for further discussion of these analyses.

Model Parameter Choices

Table 1 lists some of the parameter values used in RZWQM. Rain intensity data were obtained from a recording rainguage at Lang Farm, where the plots were located, and daily records were from a University-maintained station at Tifton, approximately 4 miles away. Soil horizon properties were obtained from unpublished data on an adjacent plot, but are quite similar to published values (12). Nine irrigation events were also accounted for.

Almost 90 cm of rainfall plus irrigation occurred during the April-Otober simulation period, and the model estimated that infiltration and runoff totaled 70 and 12 cm, respectively. Some of the precipitation occurred on the same day as a chlorothalonil application. We assumed a 20% washoff fraction (the RZWQM default for compounds in the solubility range of chlorothalonil).

Soil sorption coefficients and half-lives obtained from the the field and laboratory sampling below) were used for chlorothalonil and 4-hydroxychlorothalonil, with the degradation products of the 4-hydroxy compound being assumed to behave, initially at least, similar to their parent. Aerobic soil biodegradation was assumed to be the the major soil degradation route, with 80% of the parent conversion going to the 4-hydroxy compound, and the hydroxy compound being converted to a collection of unknown degradates whose whose sum equals the total amount of 4-hydroxy compound degraded. However, we gave these degradates a 1-year soil half-life just to keep track of them. RZWQM allows for temperature and moisture changes to affect degradation rates, and we input the conditions under which the laboratory half-lives were ovbtained: 30 ^{0}C and 0.3 cm^3/cm^3 moisture content, and used a (default) Arrhenius heat of activation of 54 kJ mol^1 ^{0}C^{-1} and Walker moisture constant of -0.8 (6). For the parent and 4-hydroxy daughter we also assumed an increase in persistence with soil septh, assuming half-lives increased linearly between 25 and 100 cm cepth up to a factor of 30 i.e., half lives of 60d for the parent and > 1 year for the degradates for depths >100 cm (7).

Table 1. Selected Parameters Used in the RZWQM Model

Parameter	Site/Crop Values, units
Field area	4.25×10^{-3} ha
Slope	1.7 %
Albedos of dry soil ,wet soil, crop	0.3, 0.15, 0.2
Evaporation Pan coefficient	0.75
Peanut planting date, seed density, depth	5/25/99, 20000 seed/ha, 2 cm
Crop height, Max. leaf area index	30 cm, 3
Maximum root depth	20 cm

Parameter (units)	Soil Horizon Values			
	Ap	Bt1	Btv1	Btv2
Type[1]	SL	SCL	SCL	SCL
Depth (cm)	0-28	28-51	51-82	82-113
Dry Bulk density (g/cm^3)	1.79	1.73	1.73	1.73
Clay (%)	0.091	0.309	0.308	0.308
Sand (%)	0.747	0.604	0.596	0.596
SHC[2] (cm/hr)	0.56	0.51	0.24	0.24
Field capacity (cm^3/cm^3)	0.18	0.20	029	0.29
Porosity (%)	32.5	34.7	34.7	34.7
Organic matter content (%)	0.6	0.6	0.4	0.24

	Pesticide or Degradate Values[3]			
	C	COH	X	Ref.
Ancestry code	P	D	G	
Priduction process	--[4]	aerobic	aerobic	
Production fraction	--	0.8	1.0	11
Molecular weight (g)	265.9	247.5	265[5]	
Aqueous solubility (mg/l)	0.6	100[4]	100[5]	15
Vapor pressure (mm Hg)	5.7×10^{-7}	0[4]	0[4]	23
Henry's law constant[6]	1.36×10^{-5}	0	0	
Foliar washoff fraction (%)	20	--	--	
Foliar washoff power paam.	0.002[6]	--	--	
Foliar deposit half-life (d)	4	--	--	19
Soil OC sorption coefficient	1600	560	500[5]	
Soil half-life	2	10	365	11
Adjust soil HL with depth?	Y	Y	N	

[1]S = sandy loam, SCL = sandy clay loam

[2]Saturated Hydraulic conductivity; crust formation assumed with SHC of 0.39 cm/hr

[3]C = chlorothalonil; C-OH = 4-hydroxchlorotyalonil; X = total products of C-OH degradation; P = parent; D = daughter; G = granddaughter code

[4]"--" = not applicable. [5]Assumed. [6]Calculated. [6]Default, based on solubility (6).

We used a simplified "generic" crop growth model provided by RZWQM and defaults for nutrient and microbiological parameters. Initial soil chemistry, plant residue, and nutrient conditions were estimated using a "wizard" provided by the model.

Chlorothalonil Foliar Deposition and Foliar Washoff by Rainfall

The foliar washoff algorithms in RZWQM had not been tested before and we found that fixes to the program were required. The program now appears to properly emulate single-application field data (e.g., 14, 20-22), but it still does not work properly for multiple applications, if the foliar persistence of the pesticide is long enough to cause buildup of foliar residues. Correcting this bug is in progress. Brenneman, et al., (18) have reported a chlorothalonil half-life in peanut foliage of about 4 days and this, combined with the considerable washoff that occurred in this experiment, means buildup of foliar residues was unlikely.

Model-estimated foliar coverage of soil, and foliar and soil surface residues of chlorothalonil are shown in Figure 3. Chlorothalonil residues approaching 0.7 kg/ha were observed by Brenneman et al., in the topmost canopy of peanuts after a 1.25 kg/ha application (18). Note that significant washoff occurred after the third and fifth applications. This, along with the first application when leaf coverage was only about 50% resulted in three major pulses of chlorothalonil to the soil.

In figure 4 the mass of the three chlorothalonil species predicted by the model in the top 2 cm of soil are compared with analyses of field soil samples taken immediately after chlorothalonil application. Major differences include (a) early field samples were actually taken between the peanut rows, where foliar interception (estimated at 60% by the model) did not occur. Thus, levels actually found were higher than predicted for the first two samples, and possibly low for the third because washoff contributions would also not be observed. (b) The last sample was taken when peanut plants were beginning to senesce, decreasing foliar interception (the model assumed that plant cover was still at the maximum). (c) Only part of the the hydroxy compound degradates were recovered, while the model assumes complete recovery.

Thus, the comparison in Figure 3 is not a good one for several reasons, but given the complexity of the processes determining the surface soil residues, it is encouraging that the data and predictions are at least on the same scale. We feel that transport predictions should be within an order of magnitude of reality.

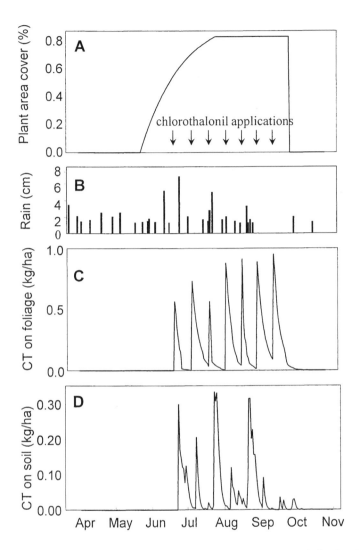

Figure 3. Model-predicted deposition, washoff and dissipation of seven broadcast chlorothalonil (CT) applications to peanuts at 14-d intervals at various stages of canopy development: (A) leaf area cover and application times, (B) rainfall amounts, (C) foliar residues, (D) soil surface residues.

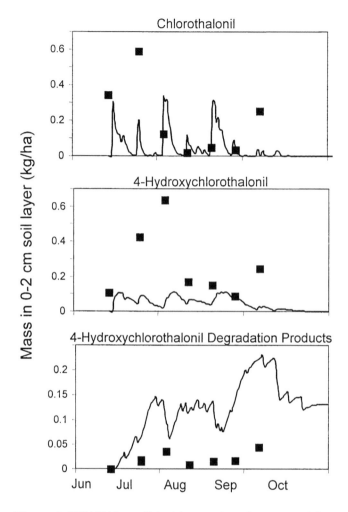

Figure 4. *RZWQM-predicted (curves) and measured (squares)
chlorothalonil and degradate amounts in the 0-2 cm layer of soil
under a developing peanut canopy. 7 broadcast applications
occurred at 2-week intervals. Note scale differences.*

Runoff and Leaching Predictions

We have only a fraction of the field data we would need to validate and calibrate the RZWMQM model to the point we would feel it's predictions for runoff and infiltration for this scenario are accurate. Nevertheless, we knowenough to make a model run interesting, if only to tell us what the major features of this system are. This is interesting because this system represents a scenario which is common in agriculture for many fungicides and insecticides--multiple foliar applications, short degradation half lives, sensitivify to washoff, and significant soil "metabolites"—but one which which has not been adequately modelled. Given the high rainfall, infiltration and runoff during the period of applications we have here a severe scenario for potential surface and ground water contamination by the parent and/or metabolite compounds.

Unfortunately RZWQM lacks an erosion model, and thus will not ordinarily give a good prediction of parent chlorothalonil runoff: with an aqueous solubility of 0.6 ug/l, we might expect that the majority of chlorothalonil will be lost in the sediment phase (24). However, in this experiment we are washing chemical off foliage and simultaneously generating runoff, and this might generate much more water-phase runoff than when the chemical is only in the soil.

This appears to be what the model run suggests. Table 2 gives runoff loads of chlorothalonil and the daughter and granddaughter(s) predicted by RZWQM. The model predicts a remarkable 12 runoff events after the chlorothalonil applications began, although 6 are very small and would likely not be seen in the field. The majority of losses are of parent chemical, in four events that occur on the day of a chemical application or shortly after. These are large losses, totaling more than 5% of what is a large amount of applied chemical. The losses in the two July events approximate the washoff fraction of the last application before runoff, and the decrease in foliar deposit can ber see in Figure 2. It is fortunate that chlorthalonil appears to be quite labile in the environment.

Leaching, on the other hand, appears to be almost nonexistent. Figure 5 shows the concentrations of the parent and degradates as a function of time and depth. The greater mobility of the degradates can be seen as well as the (false) buildup of the granddaughter products resulting from the choice for them of a very long soil half-life. In spite of their relative mobility and the low sorptivity of the soil, little movement is seen. The model predicts that all three chemicals will show a similar peak in concentration in leachate at the bottom of the root zone on Septembe 5, preceeded by smaller peaks; perhaps the soil has so little sorptivity the differences in K_{oc} have snmall effect. This peak then decreases and then increases rapidly toward the end of the simulation. All the concentrations indicted are vanishingly small, however: the largest is the granddaughter, giving a total leaching load for the period of simulation of less than 10^{-16} kg/ha, with the other compounds being much lower.

Table 2. Runoff volumes and pesticide loads for chlorothalonil , 4-
hydroxychlorothalonil

Date	Runoff	Time	Pesticide Loads (g/ha)		
		after	chlorotha-	4-hydroxy-	grand-
	(cm)	appl.[1]	onil.	chlorothalonil	daughters
6/25	0.006	2	0.4	0.0	0.0
6/28	3.385	5	56.9	8.0	2.5
7/22	0.069	13	0.7	0.1	0.3
7/23	1.024	0	164.3	0.6	2.9
7/25	2.744	1	156.1	5.5	3.7
8/1	0.07	9	0.4	0.2	0.3
8/4	0.27	12	0.4	0.4	0.8
8/11	0.16	5	9.5	0.3	0.5
8/20	0.34	0	48.0	0.5	0.8
8/21	0.01	1	0.9	0.0	0.0
8/25	0.01	5	0.2	0.0	0.0
8/30	0.04	10	0.3	0.1	0.1
TOTALS	8.128		438.2	15.7	12.0
% of applied amount[2]			5.0	0.2	0.1

[1]Time elapsed (d) between last chlorothalonil application and runoff event.

[2]4-hydroxy metabolite corrected to parent molecular weight.

PEST#1 TOTAL MASS (KG/HA)

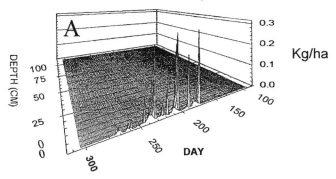

PEST#2 TOTAL MASS (KG/HA)

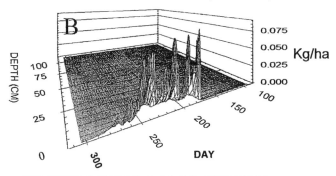

PEST#3 TOTAL MASS (KG/HA)

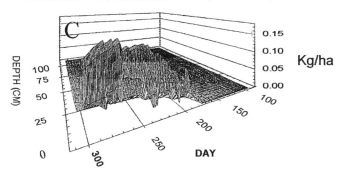

Figure 5. RZWQM post-processor generated 3-D plots of soil concentrations as a function of depth and time (Julian day) for chlorothalonil (A), 4-hydroxychlorothalonil (B), and the 4-hydroxy degradates (C). The model assumes the degradates in plot C are not degraded.

302

Acknowledgements

We acknowledge the help of Jim Hook of the University of Georgia Department of Crops and Soils for much help in obtaining weather and soil properties for the site; Chris Clegg of the ARS Southeast Watershed Research Laboratory for doing the soil sorption experiments, and Ken Rojas of the ARS Great Plains Systems Research Laboratory for essential support in using RZWQM.

References

1. Nicholls, P. H. Factors influencing entry of pesticides into soil water. *Pesticide Sci.* **1998**, 22, 123-137.
2. Beck, M.B. Water quality modeling: a review of the analysis of uncertainty. *Water Resources Res.* **1987**, 23, 1393-1442.
3. Fontaine, D. D., Havens, P. L., Blau, G. E., and Tillotson, P. M. The role of sensitivity analysis in groundwater risk modelling for pesticides. *Weed Tech.* **1992**, 6, 716-724.
4. Wauchope, R. D. Environmental risk assessment of pesticides: improving simulation model credibility. *Weed Technol.* **1992**, 6, 753-759.
5. Hanson, J. D., Ahuja, L. R., Shaffer, M. D., Rojas, K. W., DeCoursey D. G., Farahani, H., and Johnson, K. RZWQM: Simulating the effects of management on water quality and crop production. *Agricultural Systems* **1998**, 57, 161-195 .
6. Ahuja, L. R., Rojas, K. W., Hanson, J. D., Shaffer, M. D., and Ma, L.(Eds.), *Root Zone Water Quality Model: Modelling Management Effects on Water Quality and Crop Protection;* Water Resources Publictions LLC, Highlands Ranch CO., **2000**, 372 pp.
7. Wauchope, R. D., Nash, R. G., Ahuja, L. R., Rojas, K. W., Willis, G. H., McDowell, L. L., Moorman, T. B., and Ma, Q.-L. Pesticide Processes. Chapt. 7, pp 163-244 In L. R. Ahuja, J. D. Hanson, M. J. Shaffer and K. Rojas (Eds.), Root Zone Water Quality Model: Modeling Management Effects on Water quality and Production. Water Resources Publicati ons, LLC, Englewood CO 300pp. **1999**
8. Bridges, D. C., Kvien, C. K., Hook, J. E., and Stark, C. R. *An Analysis of Pesticide Use and Benefits in U.S. Grown Peanuts.* University of Georgia National Environmentally Sound Production Agriculture Laboratory. 1994. http://www.cpes.tifton.edu/nespal.
9. Culbreath, A. K., Brenneman, T. B., Reynolds, K. L., Hammond, J. M., and Padgett, G. B. Tank mix combinations of propiconazole and chlrorthalonil for control of leaf spot diseases of peanut. *Peanut Sci.* **1995**, 22, 101-105.
10. Bosch, D.D.; Sheridan, J. M.; Davis, F.M. *Trans. ASAE,* 42, 1637-1644, 1999.

11. Potter, T. L., Wauchope, R. D., and Culbreath, A. F. Accumulation and Decay of chlrothalonil and degradation products in surface soil following foliar application. In preparation.

12. Hubbard, R. K, Berdanier, C. R, Perkins, H. F., and Leonard, R. A. Characteristics of selected upland soils of the Georgia coastal plain. USDA-ARS Publication ARS-137, Oct. 1985, 70pp.

13. Ahuja, L. R., Ma, Q. Ll, Rojas, K. W., Boesten, J. J. T. I., and Farahani, H. J. A field test of root zone water quality model-pesticide and bromide behavior. *Pestic. Sci.* **1996**, 48, 101-108.

14. Willis, G, H., McDowell, L. L., Smith, S., and Southwick, L. M. Pemethrin and sulprofos washoff from cotton plants as a function of time between application and initial rainfall. *J. Environ. Qual.* **1994**, 23, 96-100.

15. Hornsby, A. G., Wauchope, R. D., Herner, A. E., *Pesticide Properties in the Environment,* Springer-Verlag, New York, NY, 1996, 227 pp.

16. van der Pas, L. J. T, Matser A. M., Boesten, J. J. T. I., and Leistra M. Behavior of metamitron and hydroxy-chlorothalonil in low-humic sandy soils. *Pestic. Sci.* **1999**, 55, 923-934.

17. Ahuja, L. R, Sharpley, A. N, Yamamoto, M., and Menzel, R. G. The depth of rainfall-runoff-soil interaction as determined by 32p. *Water Resources Res.* **1981**, 17, 969-974.

18. Ahuja, LR. Modeling soluble chemical transfer to runoff with rainfall impact as a diffusion process. *Soil Sci. Soc. Amer. J.* **1990**, 54, 312-321.

19. Brenneman, T. B., Sumner, H. R., and Harrison, G. W. Deposition and retention of chlorothalonil applied to peanut foliage: Effects of application methods, fungicide formulations and oil additives. *Peanut Science* **1990**, 17, 80-84.

20. Willis, G. H., McDowell, L. L., Meyer, L. D., and Southwick, L. M. Toxaphene washoff from cotton plants by simulated rainfall. *Trans. ASAE* **1982**, 25, 242-246,253.

21. Pick, F. E., van Dyk, L. P., and de Beer, P. R. The effect of simulated rain on deposits of some cotton pesticides. *Pesticide Science* **1984**, 15, 616-23

22. McDowell, L. L., Willis, G. H., Southwick, L. M., and Smith, S. Methyl parathion and EPN washoff from cottonplants by simulated rainfall. Environ. Sci. and Technol. **1984**, 18, 423-27.

23. Tomlin, C. T*he Pesticide Manual,* 10th Ed. Crop Protection Publications, 1994, 1341 pp.

24. Wauchope, R. D. The pesticide content of surface water draining from agricultural fields--a review. *J. Environ. Qual.* **1978**, 7, 459-472.

25. Alexander, M. Aging, bioavailability, and overestimation of risk from environmental pollutants. *Environ. Sci. Technol.,* **2000**, 34, 4259-4265.

Chapter 20

Modeling the Effect of Precision Agriculture: Pesticide Losses to Surface Waters

D. J. Mulla[1], P. Gowda[1], W. C. Koskinen[2], B. R. Khakural[1], G. Johnson[3], and P. C. Robert[1]

[1]Department of Soil, Water, and Climate, University of Minnesota, St. Paul, MN 55108
[2]Agricultural Research Service, U.S. Department of Agriculture, University of Minnesota, St. Paul, MN 55108
[3]Southern Research and Outreach Center, University of Minnesota, Waseca, MN 56093

In this study we evaluated the environmental impacts of spatially variable versus uniform applications of acetochlor. Spatially varying rates of acetochlor ranging from 2.0 to 2.7 kg ha^{-1} were applied to a 32 ha bare field planted to corn during the spring of 1998. These rates were varied in accordance with measured soil surface organic matter contents, sorption K_d values, and grassy weed populations. Surface runoff and tile drain leaching losses of acetochlor were measured using automated sampling systems. The Agricultural Drainage and Pesticide Transport (ADAPT) model was calibrated to measured water fluxes, and losses of sediment, nitrate, and acetochlor to surface waters. There was good agreement between measured and modeled water fluxes, sediment losses, nitrate losses, and acetochlor losses. Simulated acetochlor losses for the variable rate strategy were 11 - 33% lower than losses for a uniform application of 2.7 kg ha^{-1}.

Introduction

Precision agriculture is an approach that allows the rate of pesticide applied to a field to be varied in response to spatial patterns in weed populations, soil organic matter content, and pesticide sorption or dissipation characteristics. It is hypothesized that this approach allows effective weed control, maintains crop productivity (1), reduces the environmental impacts of pesticides (2), and increases farm profitability (3).

This hypothesis has not been fully tested. Stafford and Miller (4) developed a method for spraying specific patches of weeds in cereal crops. This method reduced herbicide inputs from 40-60%. Khakural et al. (5) found that variable rate applications of alachlor in small steep hillslope plots resulted in runoff losses that were as much as 24% smaller than losses from uniform applications.

The objective of this research was to evaluate long-term average edge-of-field losses of acetochlor on a commercial corn field using variable rate applications (precision agriculture) of acetochlor. These losses were compared with simulated losses of acetochlor from the same field receiving a uniform application rate of acetochlor.

Methods

A commercial corn field in Blue Earth county of southern Minnesota was selected for variable rate applications of acetochlor. Applications varied from 2.0 to 2.7 kg ha^{-1}(2). Application rates ranged from low to high label rates.

Application rates were determined using information about the spatial variability in grass weed densities and sorption partition coefficients of alachlor. A rate of 1.96 kg ha^{-1} was used with K_d values <8 ug mL^{-1} and weed densities < 4 in a 0.093 m^2 area. A rate of 2.7 kg ha^{-1} was used where K_d values were >8 and weed densities were >4. A rate of 2.21 kg ha^{-1} was applied where K_d values were <8 and weed densities were >4, while a rate of 2.45 kg ha^{-1} was applied where K_d values were >8 and weed densities were <4.

Soil organic carbon contents were determined using the Walkley Black method (6) after sampling surface soils (0-15 cm) in the field at 234 locations along 9 transects separated by 45 m spacings at intervals of 30 m within each transect. This intensive sampling strategy gave more than enough information to accurately estimate the mean, variance, and semivariograms for each of the properties measured. A subset of 42 of these samples were used to determine adsorption partition coefficients (K_d) of alachlor from standard batch equilibration techniques. Weed scouting was conducted in June of 1997 along a 15 x 15 m grid.

Topography at the site was surveyed using a geodimeter (Geotronics 126). The site has rolling topography (maximum 6% slope), with two depressional areas where water collects. Each of these depressional areas had tile drains to remove water from the subsurface and/or surface. These tile drains were equipped with automatic water quality samplers (ISCO 3700) and area velocity sensors in order to measure water flow and losses of sediment, nitrate, phosphorus, and acetochlor.

Soil types in the field were based on Blue Earth County Soil Survey Maps. Soils are typical of the region, and are primarily fine textured, poorly drained series with high organic matter content (up to 10%). Soil series included Lester, Shorewood, Cordova, Waldorf, Lura, and Blue Earth. Model inputs for these soils were obtained using the Map Unit Users File maintained by USDA-NRCS (Tables I and II).

Table I. Model input variables and their values.

Input Variable	Value
Runoff Curve Number	78
Evaporation Constant (mm $d^{-0.5}$)	4.0
Effective Rooting Depth (cm)	90
Surface Sealing Threshold (cm)	4.5
Surface Storage Depth (cm)	0.3
Depth to Impermeable Layer (m)	4.57
Initial Depth of Water Table (m)	1.25
Hydraulic Conductivity of Soil Layer (cm h^{-1})	2.54
Hydraulic Conductivity of Impeding Layer (cm h^{-1})	0.001

The Agricultural Drainage and Pesticide Transport (ADAPT) model is a combination of the surface runoff and leaching model GLEAMS (7) and the tile drainage model DRAINMOD (8). The ADAPT model runs on a daily time step, and requires input data for precipitation and temperature, cropping system (including planting and harvest dates), tillage and agrichemical management practices, slope steepness, and soil information. Soil information includes soil moisture retention characteristics, soil hydraulic parameters, and soil physical and chemical properties. Recently, the ADAPT model was calibrated and validated for continuous corn on a tile drained experimental plot located in nearby Waseca, Minnesota operated by the University of Minnesota (9). In that study, the flow and transport portions of the model were calibrated and validated using a thirteen year record of flow and nitrate loss data.

The corn field in Blue Earth county was subdivided into one hundred twenty eight 50 m x 50 m cells, and an initial set of soil and slope input factors were assigned to each cell. Sorption K_d values for each cell were obtained by cokriging interpolation of measured alachlor K_d values using measured soil organic matter as a covariate. Cokriging is known as a Best Linear Unbiased Predictor because the cokriged values at measurement locations are identical to the measured values. These cokriged sorption values were then used as surrogates for acetochlor K_d values because i) the chemical structure and properties of alachlor and acetochlor are similar, and ii) permission was denied to obtain radiolabeled acetochlor for use in the sorption experiments. The surface routing patterns for runoff and erosion from each cell were determined based on topographic slope and aspect. The subsurface routing patterns for tile drainage were determined using farmer supplied maps of tile drainage patterns.

The model was run to simulate one year for each cell, the surface and subsurface flows were routed to each of the two sample collection locations, and the predicted and measured water flows were compared. Adjustments of soil porosity, soil moisture retention, soil hydraulic conductivity, crop leaf area index, and initial depth to water table were made to improve the match between predicted and observed flows. Next, predicted and measured sediment losses were compared at the two measuring locations. Adjustments were made to the sediment delivery ratio from cells to improve this comparison. Next, predicted and measured acetochlor losses were compared. Adjustments were made to the transformation half-life of all cells to improve the agreement. The calibrated value for half-life was 19 days.

The final model parameters from this process were then used with a 3 year climatic record from nearby Waseca, Minnesota to evaluate several herbicide management scenarios over a range of climatic conditions. The scenarios evaluated were 1) a uniform application of 2.7 kg ha^{-1} acetochlor, and 2) a variable rate application of acetochlor corresponding to the rates actually applied to the field during 1998.

Table II. Values Used Layer-Specific Soil-Input Variables.

Soil Series	Soil Texture	Horizon	Thickness (cm)	POR	BR15	OM	Clay	Silt
					(cm/cm)		(%)	
Blue Earth	Mucky Silt Loam	1	25	0.61	0.37	9.34	25	66
		2	127	0.70	0.42	9.34	28	65
		3	305	0.47	0.27	0.17	25	43
Lester	Loam	1	23	0.45	0.13	3.90	21	37
		2	69	0.40	0.17	1.04	30	37
		3	366	0.38	0.14	0.39	25	37
Cordova	Clay Loam	1	36	0.51	0.19	5.50	29	37
		2	53	0.45	0.22	2.05	32	49
		3	368	0.40	0.16	0.68	24	37
Lura	Silt Clay	1	91	0.62	0.37	7.67	53	45
		2	91	0.62	0.37	0.35	44	49
		3	274	0.52	0.31	0.35	44	49
Waldorf	Silt Clay Loam	1	51	0.61	0.37	6.84	38	55
		2	64	0.53	0.32	0.68	48	47
		3	343	0.50	0.30	0.25	35	49
		1	28	0.54	0.24	5.28	35	48

Shorewood	Silt Clay Loam	2	71	0.50	0.32	1.21	46	49
		3	358	0.42	0.27	0.43	40	31
		1	28	0.54	0.24	6.56	35	48
Shorewood	Silt Clay Loam (Silty Substratum)	2	71	0.50	0.32	2.17	46	49
		3	358	0.52	27	0.77	40	31

POR = porosity; BR15 = Moisture content at wilting point; OM - Organic matter.

Results

Organic matter contents averaged 7.1% with a range from 2.2 to 10.0% (2). The highest organic matter contents were located in depressional landscapes in the northern portion of the field. The lowest organic matter contents were located along an eroded hillslope in the southern portion of the field (Figure 1).

Sorption K_d values for alachlor averaged 9.7 ug mL^{-1} with a range from 5.1 to 20.3 ug mL^{-1}(2). The highest K_d values were in a southern portion of the field with moderately high organic matter contents (Figure 2). The northern depressional region with the highest organic matter contents had relatively low values for sorption K_d.

Foxtail weed pressures in 1997 were greatest along the eastern portion of the field (Figure 3). Lowest weed pressures occurred sporadically throughout the western portion of the field, especially in the southwestern portion on steep slopes. Spatially variable applications of acetochlor were applied to the field (Figure 4) in June, 1998. The highest rate (2.7 kg ha^{-1}) was applied to the eastern edge of the field. Lowest rates (2.0 kg ha^{-1}) were applied to steep hillslopes in the southwestern portion of the field and along the western edge.

ADAPT model performance was satisfactory (Table III), with the model explaining about 60% of the measured variability, for all parameters with the exception of sediment. Performance of the ADAPT model was better for flow, nitrate, and acetochlor than for sediment (Table III). This was especially true in the northern portion of the field, where a surface tile intake collected surface runoff into a subsurface tile drainage system. The transport pathways for acetochlor include runoff, eroded sediment, and tile drain water. Good model performance in simulating flow, sediment, and nitrate improves confidence in the ability of the model to simulate all of the major herbicide transport pathways.

Most of the average acetochlor losses occur in runoff (Table IV). The average rate of herbicide applied with the uniform strategy was 2.7 kg ha^{-1}, versus 2.3 kg ha^{-1} with the variable rate strategy, about 15% less than with the uniform strategy. Total losses for the uniform strategy averaged 0.03 and 0.09 kg ha^{-1} in the northern and southern portions of the field, respectively. Losses for the variable rate strategy averaged 0.02 and 0.08 kg ha^{-1} in the northern and southern portions of the field, respectively. Total losses are 1.1% or 3.3% of the uniform rate applied in the northern or southern parts of the field, respectively.

In the flat northern portion of the field, the total losses of acetochlor with the variable application strategy are roughly 33% smaller than the losses with a uniform application strategy. In the steep southern portion of the field, the total losses of acetochlor with the variable application strategy are roughly 11% smaller than the losses with the uniform application strategy.

Table III: ADAPT Model Performance Criteria for Calibration Year (1998).

Property	Obs. Mean	Pred. Mean	R^2	Slope	Intercept	RMSE
-------------- Northern Monitoring Location --------------						
Flow (m³)	154.8	157.2	0.63	0.65	52.49	121.2
Sediment (kg)	0.9	0.7	0.27	0.38	0.59	3.6
Nitrate (kg)	2.1	1.5	0.65	0.83	0.28	1.6
Acetochlor (g)*	0.02	0.02	0.64	1.25	-0.01	0.03
-------------- Southern Monitoring Location --------------						
Flow (m³)	72.9	71.2	0.73	0.58	31.77	140.0
Sediment (kg)	1.7	1.5	0.81	0.91	0.29	4.2
Nitrate (kg)	1.0	1.0	0.57	1.26	-0.27	1.7
Acetochlor (g)*	0.01	0.01	0.57	0.46	0.00	0.01

* Acetochlor losses are for tile drainage only.

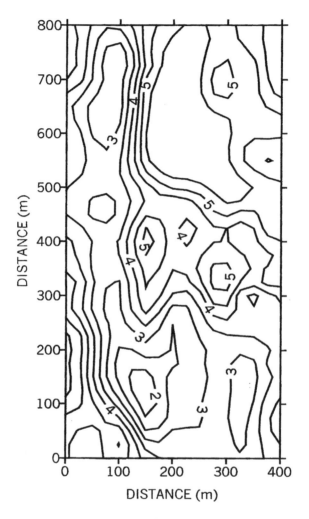

Figure 1. Spatial Pattern in Surface Organic Carbon (%).

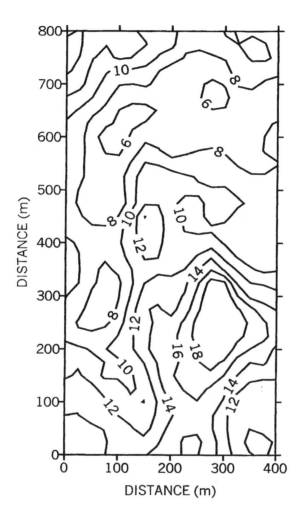

Figure 2. Cokriged Acetochlor Kd Values (ug/mL).

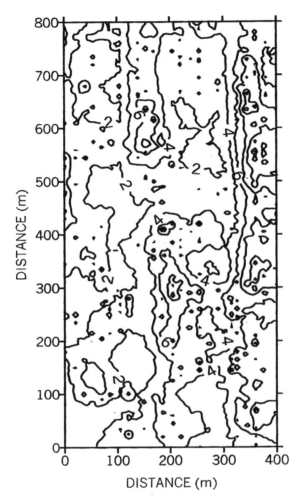

Figure 3. Spatial Patterns in Grassy Weeds
(Number per 0.304 m x 0.304 m)

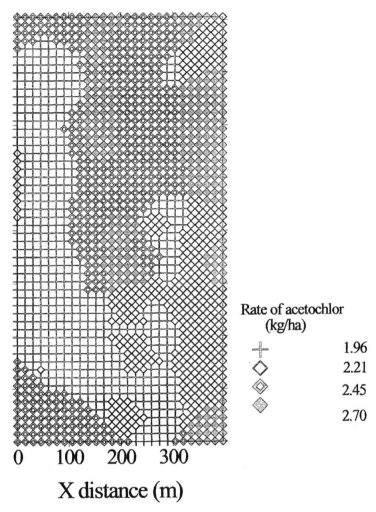

Figure 4. Site-specific Acetochlor Management Map.

Table IV. Mean Annual Acetochlor Losses (kg ha⁻¹) in Runoff, Eroded Sediment, and Drainage Water Using a 3 Year ADAPT Simulation.

Acetochlor Loss Pathways	Uniform Application Losses (kg ha⁻¹)	Variable Application Losses (kg ha⁻¹)
------------------Northern Part of Field----------------------		
Runoff	0.03	0.02
Sediment	0.00002	0.00001
Drainage	0.00003	0.00002
Total	0.03	0.02
------------------Southern Part of Field--------------------		
Runoff	0.09	0.08
Sediment	0.00005	0.00004
Drainage	0.00006	0.00004
Total	0.09	0.08

Conclusions

This is the first study to provide comparisons between field-scale losses with a variable rate versus a uniform rate application of herbicide. Modeled losses of acetochlor were roughly 1-3% of the amount applied. Losses occurred primarily through surface runoff. In the flat portion of the field, acetochlor losses were 33% smaller with the variable rate than with the uniform rate application strategy. In the steep portion of the field, losses were 11% smaller with the variable rate than the uniform rate strategy. These results indicate that measurable reductions in off-site losses of acetochlor could occur on similar sites using a variable rate application strategy in comparison to a uniform application rate strategy.

Acknowledgment

Research presented in this paper was funded by USDA-CSREES grant #95-37102-2174 entitled: "Impact of soil-specific herbicide application on water quality."

References

1. Larson, W. E., Lamb, J. A., Khakural, B. R., Ferguson, R. B., and Rehm, G. W. In *The State of Site-Specific Management for Agriculture*; Pierce, F. J. and Sadler, E. J., Eds.; Soil Sci. Soc. Am.: Madison, WI, 1997, pp 337-367.

2. Khakural, B. R., Robert, P. C., Mulla, D. J., Oliveira, R. S., Johnson, G. A., and Koskinen, W. C. In *Proc. 4th Int'l. Conf. Precision Agriculture*; Robert, P. C., Larson, W. E., and Rust, R., Eds.; Soil Sci. Soc. Am.: Madison, WI, 1999, pp 1719-1731.

3. Clay, S. A., Lems, G. J., Clay, D. E., Ellsbury, M. M., and Forcella, F. In *Proc. 4th Int'l. Conf. Precision Agriculture*; Robert, P. C., Larson, W. E., and Rust, R., Eds.; Soil Sci. Soc. Am.: Madison, WI, 1999, pp 1699-1707.

4. Stafford, J. V., and Miller, P. C. H. In *Proc. 3rd Int'l. Conf. Precision Agriculture*; Robert, P. C., Larson, W. E., and Rust, R., Eds.; Soil Sci. Soc. Am.: Madison, WI, 1996, pp 465-474.

5. Khakural, B. R., Robert, P. C., and Koskinen, W. C. *Soil Use Management.* **1995**, 10,158-164.

6. Nelson, D. W., and Sommers, L. E. In *Methods of Soil Analysis Part 2.*; Page, A. L., Ed.; Agronomy Monograph No. 9; Soil Sci. Soc. Am.: Madison, WI, 1986, pp 539-579.

7. Leonard, R. A., Knisel, W. G., and Still, D. A. *Trans. Am. Soc. Ag. Eng.* **1987**, 30, 1403-1408.

8. Chung, S. O., Ward, A. D., and Shalk, C. W. *Trans. Am. Soc. Ag. Eng.* **1992**, 35, 571-579.

9. Davis, D., Gowda, P., Mulla, D. J., and Randall, G. *J. Environ. Qual.* **2000**, 29:1568-1581.

Indexes

Author Index

Subject Index

W

Z